João Rocha
Maria Helena Carvalho Rocha
Maria Cecília Carvalho Rocha

Significant changes in the environmental landscape

João Rocha
Maria Helena Carvalho Rocha
Maria Cecília Carvalho Rocha

Significant changes in the environmental landscape

Natural phenomena or anthropogenic causes

ScienciaScripts

Imprint
Any brand names and product names mentioned in this book are subject to trademark, brand or patent protection and are trademarks or registered trademarks of their respective holders. The use of brand names, product names, common names, trade names, product descriptions etc. even without a particular marking in this work is in no way to be construed to mean that such names may be regarded as unrestricted in respect of trademark and brand protection legislation and could thus be used by anyone.

Cover image: www.ingimage.com

This book is a translation from the original published under ISBN 978-620-6-76188-4.

Publisher:
Sciencia Scripts
is a trademark of
Dodo Books Indian Ocean Ltd. and OmniScriptum S.R.L publishing group

120 High Road, East Finchley, London, N2 9ED, United Kingdom
Str. Armeneasca 28/1, office 1, Chisinau MD-2012, Republic of Moldova, Europe
Printed at: see last page
ISBN: 978-620-8-35095-6

Contents

Significant changes in the environmental landscape can be caused by both natural phenomena and anthropogenic (human) actions. Below are the differences between these two types of causes:

1. Natural Phenomena

Natural phenomena are events that occur without human intervention and can cause dramatic changes in the landscape. Some examples include:

- **Natural erosion:** Continuous action of winds, rains, rivers and seas that mould and modify the earth's surface, creating formations such as valleys, canyons and cliffs.
- **Landslides:** Displacement of large masses of soil or rock, usually caused by soil saturation, earthquakes or volcanic activity.
- **Volcanic activity:** Volcanic eruptions that can create new islands, mountains or completely modify existing areas through lava and ash.
- **Earthquakes:** Tectonic movements that can raise or lower parts of the terrain, create faults and significantly alter the relief.
- **Flooding:** The rise of water movements in rivers and seas that can deposit sediment, change river courses and submerge land.

2. Anthropic cause

Anthropogenic causes refer to changes in the landscape caused directly by human activities. Examples include

- **Urbanisation:** The construction of cities, the building of houses on hillsides and beachfronts or on cliffs, roads, dams and infrastructure that alters the relief, watercourses and vegetation cover.
- **Deforestation:** The clearing of forests for agriculture, livestock or urban development, which can lead to erosion, loss of biodiversity and changes in the local climate.
- **Mining:** Extraction of mineral resources that can create large holes, tailings piles and changes in topography.
- **Intensive agriculture:** Practices such as the excessive use of agrochemicals and monoculture can deplete the soil and change the ecological balance, leading to desertification.
- **Climate Change:** Human activities that contribute to global warming can result in melting glaciers, rising sea levels and changes in the distribution of vegetation. Changes in the environmental landscape can result from a complex interaction between natural phenomena and anthropogenic causes. While natural phenomena are part of the Earth's geological and climatic cycles, human activities have intensified and accelerated many of these changes, often with serious consequences for the environment and society.

CHAPTER 1

BUILDING DENSIFICATION ON CLIFFS: A STUDY OF THE EROSION PROCESS AT CARAPIBUS BEACH - PB, BRAZIL

The densification of buildings on cliffs can contribute to erosion discontinuity. This phenomenon occurs because the construction of heavy structures and intensive land occupation on a cliff can increase the load on the land, interfering with the natural processes of erosion and stability. Cliffs are geological formations that are naturally subject to erosion due to the action of agents such as wind, rain and seas. When the weight of buildings and soil sealing alter the natural flow of surface and groundwater, the distribution of stresses in the soil can be modified, causing slope instability. This can result in landslides and localised collapses, interrupting the natural pattern of erosion. In addition, urban densification can lead to deforestation and the loss of natural vegetation, which plays a crucial role in protecting against erosion, as plant roots help to keep the soil cohesive. Without this vegetation, the soil is more vulnerable to accelerated erosion. Therefore, excessive construction in cliff areas can aggravate erosion processes, leading to erosion discontinuity, which can cause serious problems such as soil loss and landslides, as well as threatening the safety of the buildings themselves. The aim of this research was to study the geomorphological changes that have taken place on the cliffs of Praia de Carapibus, in the municipality of Conde - Parana, and the failure to comply with the regulations established by the Environmental Law. The Google and Google Scholar websites were used to search for scientific articles in magazines and newspapers and other types of information. The cliffs, like the restinga vegetation, are considered a fragile environment and, in addition to their natural fragility, are highly sensitive due to the various activities carried out in this environment, such as urbanisation in forbidden areas and tourism. The images of houses and hotels partially destroyed by the weather on the cliffs of Carapibus beach characterise the failure of builders to comply with environmental legislation.

1 INTRODUCTION

The coastal plains of the Brazilian coast stretch between the ocean and the hills where the Atlantic rainforest is located. They have a Quaternary origin and consist of sandy marine deposits and dunes (Duarte et al., 2005).

The Coastal Zone (CZ) is an extremely dynamic area Scudelari and Freitas (p.40, 2005); Marone et al. (2012) and is characterised by being a region of ecological transition that plays an important role in the development and reproduction of various species, in addition to the genetic exchanges that take place between terrestrial and marine ecosystems. It has significant territorial overlap with the Amazon and Atlantic Forest biomes, as well as, to a lesser extent, with the Caatinga, the Cerrado and the Pampas, thus representing a complex of contiguous ecosystems forming environments of high ecological complexity.

The Brazilian coastline is an extremely diverse space due to the variety of its environmental characteristics and the socio-economic activities carried out there. Characterised as an area of multiple uses, it has over time brought together very diverse economic activities, which has led to serious land use conflicts. The diversity of natural environments and types of use of the coastal space is a complexity that requires effective planning and regulation of this occupation (ASMUS, 2004).

Brazil's coastal zone stretches from the mouth of the Oiapoque River (04°52'45"N) to the mouth of the Chw River (33°45'10"S), with a land strip of varying width stretching 10,800 kilometres along the coast and covering 17 Brazilian states, thirteen of which have their capitals on the coast, in whose metropolitan areas millions of people live, an indicator of the high level of anthropogenic pressure to which their natural resources are subjected (MMA,

2010).

The vegetation of the coastal zone is made up of flora that inhabits the region under the direct or indirect influence of the sea. Situated along the Atlantic coast, Pernambuco's coastal zone is not uniform throughout its length and its width varies from a few metres to several kilometres. Its vegetation, however, has unmistakable characteristics of its own, making it an authentic zone (LIMA, 2007). The aim of this research was to study the geomorphological changes that have taken place on the cliffs of Carapibus beach, in the municipality of Conde - Parana, and the failure to comply with the regulations established by the Environmental Legislation.

Figure 1 Coastal Zone Ecosystem
Source: Dantas, (2010)

The Coastal Zone is a region of contrasts, as it is found in areas where there is intense urbanisation, significant port and industrial activity, large-scale tourism, metropolises and regional coastal centres. On the other hand, these areas are permeated by areas of low occupation density and the occurrence of ecosystems of great environmental significance, but which have been the object of an accelerated occupation process, such as tourism, leisure and second homes (Asmus, 2004). In these situations, the common element is the diversity of the problems, the fragility of the environments and the complexity of their management, with a huge demand for training and mobilisation of the various actors involved, presupposing integrated interventions by national public policies in this region (Asmus, 2004).

The Coastal Zone shows great variations in its geological formation and, due to this differentiation as a habitat, it is home to many varieties of plants and animals. Around it are regions with a certain aridity, sandbanks and lagoons, mangroves, paludosa forests and sandy plains (Dantas et al. 2010). The first settlements were established along the coast, from which the conflicts between human activity and the environment began. Over the centuries, these conflicts have resulted in different types of disorderly occupation of the area and in production activities that damage environmental quality (Dantas et al. 2010).

Meirelles et al. (2005) systematised information on coastal processes and coastal geomorphology of a regional and local nature in order to identify and classify the forms, materials and agents of coastal dynamics.

1.1 Coastal use permits, licences and authorisations

The main regulatory framework for the use of coastal environments is environmental licensing, instituted as an instrument of the National Environmental Policy (Federal Law 6.938/81) and provided for in the 1988 Federal Constitution. It aims to introduce a preventive approach from the planning stage of intervention projects and potential activities that cause impacts on the environment (Asmus, 2004).

The Resolution of the National Environment Council - CONAMA N°237/97, which regulated and extended its scope, defines:

Environmental licensing is the administrative procedure by which the competent environmental body licenses the location, installation, expansion and operation of undertakings and activities that use environmental resources, considered to be effectively or potentially polluting or those that, in any form, may cause environmental degradation, taking into account the legal provisions, regulations and technical standards applicable to the case (FIRJAN, 2004). Environmental licensing is made up of three licences that can be issued separately or successively, depending on the nature, characteristics and phase of the undertaking or activity. These are Preliminary Licence (LP), Installation Licence (LI), Operating Licence (LO). In the context of the intense dynamics of occupation of the coastal zone, licensing as a management instrument has acted both as a corrective and a preventive measure. Licensing has a preventive nature when prior to the implementation of activities an assessment is made of the location and potential impacts, seeking to adapt the executive projects to the environmental conditions, or even not making the implementation of the enterprise feasible (FIRJAN, 2004).

1.2 Morphology of the beach environment

Beaches are dynamic systems in which natural elements such as the sea, wind and sand are in constant interaction, resulting in complex hydrodynamic and depositional processes that occur between the wave breaking zone and the post-beach region. Coastal dynamics, which determines the geomorphological construction of the coastline, is primarily responsible for the development of sandy beaches and the processes of erosion and deposition that keep them constantly changing (CUNHA, 1997).

The morphology of beach profiles in a given region can be altered due to the energetic potential of waves, since this energy is released in coastal zones (ARAUJO, 2008).

According to Santos (2007), the coastal region, as an interface between terrestrial and marine ecosystems, is responsible for a wide range of ecological functions, such as preventing flooding, salt intrusion and coastal erosion, protecting against storms, recycling nutrients and pollutants, and directly or indirectly providing habitats and resources for a variety of exploited species.

1.3 Beach vegetation

Herbaceous formations occur mainly on beach strips and dune areas, in places that can eventually be reached by higher seas, or in flooded depressions. In beach areas, frontal dunes and dunes closer to the sea, herbaceous species predominate, especially grasses, which are considered to be dune fixers (SILVA, 2009).

In the Pioneer Formations with Marine Influence, generically known as restinga, which make up the coastal region of Brazil, Assis (2011) describes the arboreal, shrubby and herbaceous types of vegetation in an attempt to include the main physiognomic variations observed from the beaches to the most inland points of the coastal plain. Anthropogenic pressure, in the sense of occupation and urbanisation of the coastal zone, has already suppressed many representative areas of this formation at various points along the Brazilian coast (Moura et al., 2012).

1.4 Topographical profile of the beach

According to Almeida (2012), greater erosion occurs due to the proximity of the beach to sandbanks and cliffs, the speed and direction of the prevailing winds, the amount of rainfall during seasonal periods, the geomorphology of the coastline and the availability and supply of sediment (Hesp, 2002).

When there is an opening in the higher areas of sand elevations, such as cliffs, the wind flow is intense and even promotes a greater opening by carrying sediment to adjacent areas (Anderson, 2006). Bernardes and Costa (2011) recorded the relative importance of edaphic and topographic gradients as environmental factors that affect the composition of physical structures in general.

1.5 Restinga

Restinga is a coastal ecosystem covered by herbaceous and shrubby vegetation located on sandy, saline terrain that can be composed of dunes and sandy plains (Rampinelli, 2011); (Sa, 2002). This biome covers around 79% of the Brazilian coast (Carvalho et al. 2001).

The flora of the restingas has some endemic species, but most of its plants can be found in other ecosystems (Sacramento et al. 2007). On beaches that have undergone intense anthropogenic action, the vegetation is restricted to low-growing plants in which grasses predominate (SILVA, 2009). Despite being distributed along the Brazilian coast, it is one of the areas most degraded by anthropogenic action over the last five centuries (ROCHA, et al. 2007).

The Brazilian Environmental Legislation CONAMA Resolution No. 303/2002 considers restinga to be a sandy deposit parallel to the coastline and generally elongated in shape, produced by sedimentation processes, where different plant communities that receive marine influence are found. They are also considered edaphic communities because they depend more on nature than on the climate. The vegetation cover on the restingas occurs in a mosaic and is found on beaches, sandy ridges, dunes and depressions, presenting, according to the stage of succession, a herbaceous, shrub and tree layer, the latter being more inland (MMA, 2002).

Dating back to the Quaternary period, sandbanks have a recent sedimentary origin, but the species of fauna and flora that live there have mechanisms to withstand the dominant physical factors, such as salinity, strong winds, little water, high temperatures and soil instability (ALMEIDA, 2012).

According to Carvalho et al. (2012), the process of restinga formation began in the Quaternary when the sea retreated, and they are still undergoing a dynamic process of formation and modification. The vegetation coexists with high salinity, exposure to the sun and winds (Silva et al. 2009).

1.6 Environmental impact

The restingas with their vegetation, riparian forests, lakes, dunes and sea species were incorporated into the permanent preservation areas (APP) by the Forestry Code (Law 4.771/65-Brazil/1965), with the aim of preventing their de-characterisation through anthropic action because they have an important ecological function in nature, mainly because they maintain the integrity of the most fragile areas (Santos, 2007). Their high diversity indicates the need to protect as many environmental situations as possible (Sztutman and Rodrigues, 2002). These areas harbour groups of tree species differentiated by highly evolved niches (Denslow, 1987). The sandbanks that fix dunes and cliffs have been one of the most compromised areas, often with the consent of environmental agencies that authorise the construction of homes, hotels and leisure areas on them. The Brazilian government has also paid special attention to the sustainable use of coastal resources in order to organise the occupation of coastal areas and, to this end, has created the National Coastal Management

Plan (PNGC), implementing a process marked by experimentation and constant improvement (Santos, 2007). Compromise of the coastal vegetation cover, such as the restinga vegetation formation, is one of the areas that has been de-characterised in order to meet the demand arising from occupation aimed at making a profit, in disagreement with environmental legislation and due to the inefficiency of environmental agencies, added to the social and economic crisis, and the absence of an effective environmental policy that end up compromising the biome (Santos, 2001). While the environmental recovery of dune cordons is already a common practice, cliff areas have not yet received the attention they deserve. Various recovery projects have been carried out and it is possible to recover compromised areas, as long as the tensile forces that prevent their recovery are removed (Rampinelli, 2011). The dynamic study of the natural regeneration of disturbed sandbank areas in their different communities is an interesting strategy for understanding the processes and how long they work in such a complex environment (Sa, 2002). Urbanisation leads to a loss of above-ground biomass, promoting variation between landscapes (Nascimento et al. 2002); (Alves et al. 2010). Fachelli and Pickett (1991) state that the production of rubbish can produce important variations in the environment.

Article 1 of the National Environmental Council (CONAMA) Resolution of 23/01/1986 defines Environmental Impact as:

Any alteration of the physical, chemical and biological properties of the environment caused by any form of material or energy resulting from human activities that directly or indirectly affects health, safety, social and economic activities, biota, the aesthetic and sanitary conditions of the environment and the quality of environmental resources.

The urbanisation of coastal areas occurs for leisure purposes with the establishment of temporary housing, high standard condominiums or buildings, generally also occupying areas of sandbank. According to Ackerly (2003), ecological triage is important in community assembly, migration and habitat monitoring so as not to disfigure the landscape. The most important consequences of this occupation are the elimination of natural vegetation (Malhi et al. (1999), the stimulation of erosion processes, changes in drainage characteristics due to cuts and embankments, which require borrowed material obtained from the excavation of hills located on the coastal plain, the generation of waste, the generation of domestic sewage, generally without adequate treatment and drainage problems due to the outcropping of phreatic lengol on the flat areas of the coast, as well as an increase in the demand for natural resources (Montovani, 2000). Uncontrolled urban development and pollution are not appropriate for these coastal ecosystems and are the main threat to the Brazilian tropical coast (Leao and Dominguez, 2000).

In an attempt to contain the degradation of sandbanks and cliffs, the Brazilian Forestry Code, through Law 4.771 of 15 September 1965, classifies them as a Permanent Preservation Area (APP), which cannot be devastated, according to Article 2, point "f". Conama Resolution 303, of 20 March 2002, which sets out the parameters, definitions and limits of APPs, establishes that a Permanent Preservation Area is the area located in the regions of cliffs and sandbanks: in a minimum strip of 300m measured from the line of maximum high tide; or in any location or extension, when covered by vegetation with the function of fixing these areas.

1.7 Falesias

The definition of phalesia used by Ab'Saber (2005) corresponds to an abrupt wall caused by marine erosion in front of coastal points or promontories. Christofolett (1980, p.133) defines phalesia as a ledge not covered by vegetation, with very steep slopes and varying heights, located on the coastline where the land meets the sea. The retreat of the falesia towards the mainland expands the surface eroded by the waves, forming the so-called abrasion terrages. Cliffs in Brazil occur mainly in the northeast and southeast regions, where they alternate with beaches, dunes, mangroves, reefs, bays and sandbanks, giving the coastal landscape its uniqueness (Assad, 2010).

They are considered fragile environments and, in addition to their natural fragility, cliffs are highly sensitive due to the various activities carried out in this environment, such as tourism, housing, industry and commercial spaces (Muehe and Nicolodi, 2008).

Suguio (1998) defines cliffs as steep-sided cliffs formed by the erosive action of waves on rocks. Silva (2004) states that a coast with cliffs made of consolidated material (igneous rock) can have an erosion rate of around one metre per year, while cliffs made of semi-consolidated material (sedimentary rock) can have rates of around one centimetre per year, or even metres per year due to episodic events. Developments on the edges of cliffs alter the topography of the region and deforestation accelerates the process of rain erosion, interfering with the natural retreat of the coastline (Scudelari and Freire, 2005).

One of the serious problems related to the expansion of local tourism is the occupation of the edge of the cliffs, which, together with coastal erosion, has caused a number of problems for landowners, public authorities and nature (Freitas Neto et al. (2005, p.2067-8).

Due to their porosity, semi-consolidated cliffs can temporarily store rainwater or surface runoff and subsequently open cracks in their escarpment, forming spouts in the steep part of their morphology, which can be compromised by urban occupation. Dead trees make it easier for rainwater to run off, causing barriers with large volumes of sediment to collapse over time (Pinheiro, 2003).

2 METHODOLOGY

Google and Google Scholar were used as search engines to obtain data on articles from the Scielo online library, dissertations, theses, blogs, among others to support this work. Field images were taken by the author and Google Earth images were used to record the urban invasion of the Carapibus cliff area, from hotels, inns and houses, as well as comparing it with the neighbouring Tabatinga beach, which does not allow urbanisation on the cliffs due to its highly rugged terrain.

2.1 Characterisation of the study area

The coast is subject to natural stresses such as wave action, intense rainfall, salinity, winds and steep slopes, which cause constant physical changes in the region, making it an area with a high degree of geomorphological instability. This factor is increasingly aggravated by unrestrained land occupation, due to property expansion that attracts developments and large flows of people to a region that cannot withstand these developments. In the municipality of Conde, located on the south coast of Paraná, these urban occupation projects can be observed on the cliffs, especially on Carapibus beach.

Inns, hotels and holiday condominiums are some of the most common developments occupying the cliffs. The south coast of Paraná is made up of a geological belt characterised by the deposition of sediments in Barreiras, which is described as a sedimentary cover of medium and coarse continental sandstones, as well as conglomerate sandstones over carbonate sandstone deposits (Barbosa et al. 2003). In this way, the environment in which the population promotes settlement can have its natural dynamics altered due to geothermal risk situations that may arise, which is why the focus is on the danger that these situations offer (Rosseti et al. 2009).

2.2 Research Area Description

Carapibus beach is located in the municipality of Conde, in the metropolitan area of Joao Pessoa, on the south coast of Paraná - Brazil, and has a large population in the coastal area, especially near the cliffs. The municipality has approximately 24,000 inhabitants (CPRM, 2005). The cliffs, which are present on almost all of the municipality's beaches, vary in height from 12 metres to 50 metres. The upper part of the cliffs corresponds to a threshold for the physiographic unit of coastal trays that extend throughout the municipality (CPRM, 2005). Carapibus beach has a very narrow strip of sand between the sea and the base of the cliffs.

Several houses and some inns, houses and hotels have been compromised by the collapse of the soil located on the edge of the cliff, which is suffering the consequences of the weathering of its structure, which is gradually moving massively, causing part of the buildings to collapse (CPRM, 2005).

3 RESULTS AND DISCUSSION

Figure 2 shows an image of one of the first residences in the vicinity of the site where the Brisas de Carapibus hotel was opened, located on the upper part of the cliff, with 31 flats belonging to the British J L Group. The arrow shows part of the facilities built on the edge of the cliff. Unfortunately, the source of this information did not give the date of the photo. At that time, the beach had positive sediment and sailing boats could move about freely.

Figura 2 Record of the First Building on the Falesia of Carapibus
Source: http://blog.tribunadonorte.com.br

Figure 3 shows two four-storey high buildings built on the cliffs of Carapibus beach, approximately 70 metres from the sand. It was observed that other pousadas, as well as holiday homes, have been built nearby, not observing the distance established by the supervisory bodies.

Figura 3 Medium-Sized Constructions on the Carapibus Cliffs
Source: Joao Vitaliano - July/2018

Figure 4 shows the attempt to prevent the waves from crashing into the base of the Carapibus cliffs, in which small retaining walls were built at some points considered critical. Some trees that didn't belong to the restinga flora and were part of the garden of the houses gave way and slid along with the sediment to the beach. The construction of the retaining walls prevented the passage of vehicles that constantly travelled this route with tourists.

Figura 4 Retaining Walls to Prevent Wave Shock
Source: Joao Vitaliano - July/2018

Figure 4.1 Retaining Walls Invaded by Waves
Source: Joao Vitaliano - December/2019

The image in Figure 5 shows the wear and tear of the cliff, whose opening caused by the rains has completely destroyed the stairs leading from this house to the beach. Part of the wall with the vegetation, especially the coconut palms, has completely collapsed. The wear and tear caused by the rains was so intense that it is possible to see, from the arrow, the compact block of sediment left over from the cliff structure that once existed at this location.

Figura 5 Wear and tear of the cliff with destruction of the wall and staircase
Source: Joao Vitaliano - July/2018

Figure 6 shows the critical situation that urbanisation on a cliff faces every year when the rains are heavier, regardless of whether the season is rainy or not. Arrow number one shows a house on the verge of collapsing after the sediment block has completely collapsed, leaving its base completely exposed. Arrow number two shows the slumped material containing parts of the structure that was attached to the wall. Arrow number three shows the oldest block of sediment that withstood the wear and tear, remaining eight metres away from the barrier that collapsed in the rain. Arrow number four shows the floor of the limestone on which the structures were built. The other structures on this site are no more than ten metres from the edge of this cliff.

Figura 6 Evidence of Structural Wear in the Upper Area of Falesia

Source: Joao Vitaliano - July/2018

Figure 7 shows the destruction of the cliff through a fault excavated by the rains over time, in which a staircase was built for residents to access the beach. The opening occurred due to the strong flow of water promoting slipping.

Figura 7 Falling Wall and Complete Destruction of Staircase

Source: Joao Vitaliano - July/2018

Figure 8 shows the extensive destruction of the walls and part of the house located on the edge of the cliff. Rainfall is considered by local residents to be the main factor responsible for accelerating the wear and tear of this area.

Figura 8 Accentuated Wear and Tear of Residence in Alto da Falesia
Source: Joao Vitaliano - July/2018

Figure 9 shows some structures on the verge of collapsing completely. It can be seen that the waves at high tide, except at sizfgia tide, do not reach the base of the cliff.

Figura 9 Structures and Houses on the Verge of Collapsing
Source: Joao Vitaliano - July/2018

Figures 10 and 11 show other structures affected by the recent rains that left all the residents of this area of Carapibus beach without access to the seafront after the staircase was destroyed. There has been a drop in tourism in this area because the inns and hotels built in this perimeter have been seriously jeopardised.

Figures 10 and 11 Compromised Construction and Positive Sediment
Source: Joao Vitaliano-July/2018

Figure 12 and 13 show the beginning of the wear and tear process on the cliff structure. The uncompacted soil is yielding to weathering, wearing down the structure.

Figure 12 Wear and tear on the structure Figure 13 Palm tree falling from the garden
Source: Joao Vitaliano - July/2018

Figures 14 and 15 show the risks of large-scale construction on the surface of fragile structures such as cliffs. A lot of money is invested in these works without a thorough examination of the risks posed by the environment. Figure 14 shows compact blocks of sediment that have collapsed under the action of this year's rains. In Figure 15, an image of undesigned urbanisation encroaching on the top of the cliffs at Carapibus beach.

Figure 14 Falling Blocks and Positive Sediment Figure 15 Urbanisation without a Project
Source: Joao Vitaliano - July/2018

Figure 16 shows the collapse of a huge block that was part of the cliff with its Restinga vegetation, exposing the structures to the risk of collapse. Figure 17 shows the image taken from the roof of the building shown in Figure 2.

Figure 16 Sliding of the cliff into a block Figure 17 Urbanisation on the cliffs

Source: Joao Vitaliano - July/2018

Figure 18 shows the image of Carapibus beach in December 2019, an area in which a large part of the vegetation located on the edge of Falesia was projected onto the beach due to the rains that occurred between May and June.

Figura 18 Projection of Vegetation in the Upper Area of the Falesia

Source: Joao Vitaliano - Dec/2019

Figure 19 shows the block fall of large areas of the Falesia with vegetation in the lower part of the image. The structures of the buildings are gradually appearing. In this case, the clay soil has caved in due to the infiltration of rainwater through the roots of medium and large trees.

Figura 19 Compromised Structures in the Upper Area of the Phalisis
Source: Joao Vitaliano - Dec/2019

Figure 20 shows a huge number of blocks that gave way after the rains in May and June, as well as the sediment that became negative, preventing vehicles and people from travelling on the high tide.

Figura 20 The Carapibus waterfront is blocked
Source: Joao Vitaliano - Dec/2019

4 FINAL CONSIDERATIONS

Some important points observed in this study are in line with the thinking of the authors cited in relation to the fragility of the cliffs, urbanisation projects keeping a regular distance from the beach established by the competent bodies, preservation of these environments as Permanent Preservation Areas, protection of the restinga vegetation that protects the soil and greater awareness among inhabitants not to take up residence in coastal environments without a prior study of the interactions between the soil and the climate of a given region, as well as the probable risks that characterise the dynamics of the systems under study. It is important to understand the interaction, changes and responses that will occur in coastal environments in relation to the sedimentary balance when hit by the weather, especially the coastline, which absorbs the impact of salinity, winds, storms and rains. The serious problems faced by the houses located on Carapibus beach today are due to urbanisation without a Master Plan, through which it would be possible to implement an environmental policy in line with sustainability standards. We were at Carapibus beach once again this year (2022) in January and the fall of the cliffs shown in the images from previous years continues to occur, jeopardising several stretches not previously shown.

5 REFERENCES

AB'Saber, A. litoral do Brasil/Brazilian coast. Sao Paulo. Metalivros. Translated by Charles Holmquist 2005. 281p.

Ackerly, D D. 2003. Community assembly, niche conservatism, and adaptive evolution in changing environments. Int.J.Plant Sci. 164(3Suppl0:165-184. http://dx.doi.Org/10.1086/368401

Almeida, J R. Suguio, K. Potencialidade Geoturistica das Dunas Eolicas da Ilha Comprida. Geosciences. V.31, n.3, pp.473-484. 2012.

Alves, L F. Vieira, S A. Scaranello, M A. Camargo, P B. Santos, F A M. Joly, C A. Martinelli, L A. 2010. Forest structure and live aboveground biomass variation along an elevational gradient of tropical Atlantic moist forest (Brazil). Forest Ecol. Manag. 260:679-691. http://dx.doi.org/10.1016/j.foreco.2010.05.023

Anderson, J L. Walker, I J. Airflow and sand transport variations within a backshoreparabolic dune plain comple: NE Grahan Island, British Columbia, Canada. Geomorphology. 77.17-34. 2006.

Araujo, R S. Morphology of the beach environment, sedimentology and historical evolution of the coastline of the beaches of Itapocoroi cove - Santa Catarina. University of Vale do Itajar Dissertation. 2008. 143p.

Asmus, M L. Kitzmann, D I. Coastal Management Laboratory. Coastal Management in Brazil. Current Status and Perspectives. Montevideo. 2004. 63p.

Assis, Marco Antonio et al. Restinga and lowland forests in the coastal plain of southeastern Brazil: vegetation and environmental heterogeneity. Biota Neotrop. 2011. Vol.11, n.2, pp.103-121. 'http://dx.doi.org/10.1590/S1676-06032011000200012.

Barbosa, J A. Souza, E M de. Lima Filho, M F. Neumann, V H. The stratigraphy of the Parana Basin: a reconsideration. In: Neves, Benajamim, Bley de Brito et al. New geological and geophysical data for the geometric and stratigraphic characterisation of the Alhandra sub-basin. Geol. USP, Ser. Cient., Sao Paulo, v.9,n.2, pp.63-87, June. 2009.

Bernardes, Carolina. Costa, Flavia Regina Capellotto. Effects of environmental variables on the composition of Piper assemblages at the mesoscale in the Madeira-Purus interfluve. Biota Neotrop. Vol.11, n.3, Campinas, July/Sept. 2011. http://dx.doi.org/10.1590/S1676-06032011000300006

Carvalho, L C. Freitas, A F N. Rocha, C F D. Sluys, M V..Variation in the structure and composition of Bromeliaceae in five restinga zones in the Jurubatiba Restinga National Park, Macae, RJ. Ver.bras.Bot. vol.24, n.1. Sao Paulo. Mar. 2001.

Christofoletti, Antonio. Geomorphology. Sao Paulo, Edgar Blucher, 2ª edigao. 1980. 188p.

CPRM - Servigos Geologicos do Brasil (2005). Project to register underground water supply sources. Diagnosis of the municipality of Conde, state of Parafoa. Beltrao, B A. Morais , F de.Mascarenhas, J de C. Miranda, J L F. Souza Junior, L C. Mendes, V A. Available at: http://www.cprm.gov.br/publique/?tpl=home . Accessed at 00:10hs.

Cunha, fcaro. Sustainable Development on the Brazilian Coast. Galician Journal of Economics. Vol.14, n.1-2 . 2005.pp. 1-14.

Dantas, L P. Pompei, N C. Guido, T S S. Ressuti, W. Replanting of Pau-de-Balsa and its Environmental Impacts. Sao Paulo. 2010. 69p.

Denslow, J S. 1987. Tropical rainforest gaps and tree species diversity. Ann.Rev.Ecol. Syst. 18:431-451. http://dx.doi.org/10.1146/an n u rev. es.18.110187.002243

Duarte, Heitor Monteiro. Geblerd, Arthur. Scaranoc, Fabio Rubio et al. Ecophysiology of six selected shrub species in different plant communities at the periphery of the Atlantic Forest of SE-Brazil. Flora.200.pp.456-476. 2005. doi:10.1016/j.flora.2005.02.004

Facelli, J M. Pickett, S T A. 1991. Plant Litter: its dynamics and effects on plant community structure. Bot. Rve.57(1):1-32. http://dx.doi. org/10.1007/BF02858763

Firjan/2004. Environmental Licensing Manual: step-by-step procedural guide. Rio de Janeiro: GMA, 2004. 293p.

Freitas Neto, Osvaldo, et al. Study of the dynamics of cliffs in the municipality of Tibau do Sul - RN. 1st Brazilian Congress of Environmental Engineering Geology ABGE, 13 to 16 November 2005. Florianopolis, SC.

Hesp, P. Foredunes and blowouts: initiation, geomorphology and dynamics. Geomorphology. 48. 245-268. 2002.

Leao, Z M A N. Dominguez, J M L. 2000. Tropical coast Brazil. Mas. Poll. Bull. 41:112122. http://dx.doi.org/10.1016/S0025- 326X(00)00105-3

Lima , D A. Phytogeographical studies of Pernambuco. Anais da Academia pernambucana de Ciencias Agronomicas, vol.4, pp.243-274. 2007.

Marone, e.; Mantovanelli, A.; Noernberg, M A.; Klingenfuss, M S. Lautert, L F C.; Prata Junior, V P. Physical characterisation of the Paranagua Bay estuarine complex. In: Lamour, M R. Odreski, L L R. Soares, C R. Considerations Regarding Shoreline Morphology Variation at an inlet in Southern Brazil. Journal of coastal research. SI.39, pp.565-567. 2004.

Meireles, Antonio Jeovah de Andrade. Arruda, Maira Gomes Cartaxo. Gorayeb, Adryane. Integration of geo-environmental indicators of relative sea level fluctuations and climate change on the Ceará coast. Mercator. Revista.Geogr.UFCE, ano 4, n.08. 2005. **DOI 10.4215/RM0000.0000.0000**

MMA. Management of Aquatic Biodiversity and Fisheries Resources. Overview of the conservation of Coastal and Marine Ecosystems in Brazil. Brasilia. MMA/SBF/GBA. 2010. 148p.

MMA/CONAMA/2002 - Restinga VIII. Available at: http://www.mma.gov.br/port/conama/res/res02/res30302.html. Accessed on: 11/08/2018 at 13:26 hs.

Malhi, Y., Baldocchi, D d.; Jarvis, P G. 1999. The carbon balance of tropical, temperate and boreal forests. Plant cell Environ, 22(6):715-740. . http://dx.doi.org/10.1046/j.1365-3040.1999.00453.x

Mantovani, *W. 2000. The* coastal region of São Paulo. Pp.33-41. In: Guedes, Daniela; Barbosa, Luiz Mauro. ; Martins, Suzana Ehlin. Floristic composition and phytosociological structure of two fragments of restinga forest in the municipality of Bertioga, SP, Brazil. Acta bot.bras. 20(2): 299-311. 2006.

Moura, L T L. Velozo, T M G. Borba, C S. Gomes, S M. Botanical List and Impacts caused by Trails in the Sandy Cordon in the Barra do Rio Mamanguape-PB Apa. XI Brazilian Congress of Ecology, September, Porto Seguro - BA. 2013.

Muehe, Dieter. Nicolodi, Joao Luiz. Geomorphology. In: Macro diagnostico da zona costeira e marinha do Brasil. Brasília. MMA. 2008.

Nascimento, H EM. Lawrence, W F. 2002. Total above ground biomass in central Amazonian rainforests: a landscape-scale study. Forest Ecol Manag. 168(1-3):311-321. http://dx.doi.org/10.1016/S0378- 1127(01)00749-6

Rampinelli, I T. Vegetation Structure in Areas of Permanent Preservation (APP) in Restinga Ecosystems, Southern Santa Catarina. Monograph. 2011. 42p.

Rocha, Roberto Antonio. 2007. Costa do Conde, on the coast of Paraíba, opens a hotel next Saturday. Available at: http://blog.tribunadonorte.com.br/eturismo/date/2011/04/page/6 . Accessed on 09/08/2018 at 15:41.

Rosseti, D F.; Almeida, S.; Amaral, D D.; Lima, c m.; Pessenda, L C. 2009. Coexistence of forest and savanna in an Amazonian area from a geologycal perspective. Journal of Vegetation Science. 21:120-132.

Sa, Cyl Frney Catarino de. Regeneration of a stretch of restinga forest in the Jacarepia State Ecological Reserve, Saquarema, State of Rio de Janeiro: II -Shrub substrate.

Rodriguesia,53(82):5-23. 2002.
Santos, C R. Proposal for planning criteria for the area of direct influence on the permanent preservation area (dune-fixing restinga vegetation) of the Santa Catarina coast: bases for integrated coastal management. Journal of integrated coastal management. 7(1):17-32.2007.
Santos, C R. The interface between policies and the process of human occupation in the area of permanent preservation: dune-fixing vegetation on the island of Santa Catarina, SC. Thesis. UFSC. Florianopolis. 388p. 2001.
Scudelari, A C. Freire, L C S. Determination of Environmental Impacts in the Construction of a Hotel System on the Falesias in Tibau do Sul - RN. Braz.J.Aquat.Sci.Technol. 2005. 9(1):39-43.
Sztutman, M. Rodrigues, R R. 2002. The vegetation mosaic in an area of continuous coastal plain forest, Campina do Encantado State Park, Paripueira-Agu, SP. Ver.Bras.Bot. 25:161-176. http://dx.doi.org/10.1590/S0100-84042002000200005
Silva, A N. Silva, A F. Neto, A D S. Bioma Restinga.Aula/Slides. Technology Centre - Ecology. UFAL. 2009. https://pt.slideshare.net/JadeCorte/apresentao-restinga. Accessed on 11/08/2018 at 12:16 pm.
Silva, Cleverson Guizan. Coastal sedimentation environments and morphodynamic processes acting on the coastline. In: Baptista Neto, J A. Ponzi, V R A. Sichel, E S R. Introduction to Marine Geology. Rio de Janeiro. Interciencia. 2004.
Souza, M P. Ferreira, J C P. Melo, M T C. Sedimentological characterisation of the Cacimbinhas Dune - Tibau do Sul - RN. Revista Geonorte, V.1, N.4, p.526 - 539. 2012.
Suguio, Kenitiro. Dicionario de Geologia sedimentar e áreas afins. Rio de Janeiro. Bertrand Brasil. 1998. 1190p.

CHAPTER 2

CONSEQUENCES OF CONSTRUCTION IN THE POST-BEACH ZONE: A STUDY OF EROSION ON CAPONGA BEACH - CE, BRAZIL

Construction in the post-beach zone, which is the area located immediately after the strip of sand, can have a number of environmental, social and economic consequences. Some of the main consequences are listed below:

1. Coastal erosion

The construction of buildings in the post-beach zone can intensify coastal erosion processes. Structures such as buildings, roads and walls alter the natural dynamics of wind and water currents, often interrupting the flow of sediment that naturally replenishes the sand strip. This can lead to a reduction in the width of beaches and the advance of the sea, putting the buildings themselves and the integrity of the coast at risk.

2. Ecosystem destruction

The post-beach zone is often home to sensitive ecosystems such as dunes and mangroves. Building on these sites can lead to the destruction of these areas, which are important for biodiversity, acting as breeding sites for various species and acting as natural barriers against erosion and storms.

3. Increased Risk of Flooding

Urbanisation in the post-beach zone can reduce the infiltration capacity of the soil, since sealing the surface with asphalt, concrete and other constructions prevents rainwater from being absorbed. This increases the risk of flooding, both in the built-up area and in adjacent regions, especially during storms and high seas.

4. Loss of Public Access to the Beach

Construction in the post-beach zone, such as resorts, hotels and private condominiums, can limit or prevent public access to beaches. This is particularly problematic in areas where the beach is a natural resource of great cultural and recreational value to the local community. Privatising access to beaches can cause social conflicts and reduce the well-being of the population.

5. Visual and Aesthetic Impacts

Construction in post-beach areas can significantly alter the coastal landscape, often in a negative way. The presence of high-rise buildings and infrastructure can destroy the natural view of the coastline, impacting tourism, which often depends on the scenic beauty of the beaches.

6. Increased Vulnerability to Natural Disasters

Buildings in the post-beach zone are particularly vulnerable to natural disasters such as tsunamis, hurricanes and tropical storms. As well as putting human lives and material goods at risk, frequent reconstruction after these events can be economically unsustainable and environmentally damaging.

7. Changes in Social Dynamics

Excessive urbanisation of the post-beach area can lead to gentrification, where property development drives out traditional residents and replaces the local population with tourists and new residents with greater purchasing power. This can result in a loss of cultural identity and a change in local social dynamics.

Construction in the post-beach zone, although often driven by economic and tourist development, can have a number of negative consequences for the environment and local communities. It is crucial that urban planning in these areas takes environmental preservation, public access and risk mitigation into account in order to guarantee sustainable and balanced development.

The urbanisation of the coastal zone is a global phenomenon. The absence of adequate planning with limits to protect the shoreline has resulted in disorderly occupation along practically the entire beach arc, with buildings positioned close to the active beach profile. Anthropogenic intervention in coastal processes through urbanisation has been taking place on several beaches along the Brazilian coast, leaving building constructions vulnerable within the beach's dynamic response range to storms and high seas, with the tendency for the built-up area to be reclaimed by the sea. The aim of this research was to study the phenomenon of coastal erosion on Caponga-CE beach and the consequences of high seas.

1 INTRODUCTION

The rise in the global mean sea level is related to the Earth's natural climate variability, mainly due to the thermal expansion of the ocean, caused by the increase in the volume of water induced by the rise in atmospheric temperature. The decrease in the amount of sediment in certain areas can be explained by the many human activities in riverside or inland areas, including dams, dredging and aggregate extraction.

Mendes (2008) considers that the destruction of natural structures is due to the anthropogenic degradation of the resulting natural coastal forms induced by their occupation with constructions. According to the same author, heavy engineering works create local imbalances, causing disturbances in the coastal stretches in which they are inserted: transverse structures - spurs interrupt the littoral transit of sand and cause accumulation of sand on one side of the spur and deficit on the other; longitudinal structures have the main consequence of inhibiting exchanges between dunes and beaches.

According to the Ministry of the Environment (2008), among the economic activities that are considered most dangerous when it comes to the use of beach areas is the proliferation of built-up fronts. Anthropogenic action threatens the ability of coastal zones to adapt to the effects of climate change, particularly meteorological seas and the increased frequency and intensity of storm events.

According to the same author, anthropogenic intervention in coastal processes, followed by urbanisation of the shore, has been identified as the main cause of erosion on various beaches along the Brazilian coast, mainly because the construction of buildings is vulnerable within the beach's dynamic response range to storms and high seas, which tends to lead to the built-up area being taken over by the sea.

Souza et al, (2003) points to the urbanisation of the coastline, sealing of marine terrain and occupation of the post-beach area as the main anthropogenic causes of coastal erosion.

Indicators of coastal erosion include a very narrow or non-existent post-beach due to flooding by high tides and the destruction of dunes with restinga vegetation on urbanised or non-urbanised beaches. The action of the sea plays an important role in coastal morphodynamics and hydrodynamics, mainly in terms of extending the wave action zone and generating currents in estuaries and shallow waters, which is important for sediment transport (Souza et al, 2003).

Morais (1980) observed the behaviour of the seas in the Porto de Mucuripe basin and determined a maximum amplitude of 2.7m for the equinox of margo in 1976 and 3.3m for maximum siz^gia in 1980.

Maia (1998), using 14 monthly analog records from the tide gauge installed at the Port of Mucuripe, found that during the period from May to June 1996, the maximum amplitude of the tide was 3.23 metres, at high tide in December 1995. According to the same author, in the urban centre of Caponga beach there was a certain frequency of a 50° direction in the break, this angle being the result of the diffraction of the waves when they collide with the reefs in the area, resulting in a higher percentage of frontal attack on this stretch of beach.

2 METHODOLOGY

This study used the method of obtaining images obtained by suborbital and ground remote sensing to compare the state of erosion observed on Caponga beach in the state of Ceará.

Images were taken using the method based on the concept of the equiKbrium profile, which translates the shape of the beach for a given state of constant agitation and for a given granulometry of the sand. According to Morais (1980), analysing the morphodynamic state of a given area consists of comparing the current profile with the previously observed profile, allowing us to conclude whether the stretch is eroding or accreting. The ground images were taken during the first weeks of January 2013 and January 2014, when the tidal waves provided amplitudes of 2.8 to 3.0 metres, reaching previously unrecorded locations.

3 RESULTS AND DISCUSSION

The images in Figures 1, 2, 3 and 4 show how great the sediment dynamics are on Caponga beach, caused mainly by the high seas recorded in 2013. Morais (1980) states that wave action at this location is softened by the formation of reefs that promote great diffraction, but even so, there is a very strong scraping of the sediments. It is important to emphasise that this restaurant area was built within the post-beach area, as can be seen in the suborbital image obtained from Google Earth.

Figure 1 Positive Sediment/Scale Figure 2 Negative Sediment/Scale

Photo: Joao Vitaliano. Jan/2013 Photo: Joao Vitaliano. Jan/2014

Figure 3 Positive Sediment/2013 Figure 4 Negative Sediment/2014

Photo: João Vitaliano. Jan/2013 Photo: João Vitaliano. Jan/2014

In the beginning of the second week of January 2013, high seas began to scrape away the sediment that had accumulated in front of the restaurants, as can be seen in Figure 5. In the first days of January 2014, we recorded the action of the sizfgia tide with waves that invaded the houses and reached the second block of flats, a moment we recorded on video and uploaded to YouTube.

Figure 5 Tidal Range 2.6 Figure 6 Tidal Range 3.0
Photo: João Vitaliano. Jan/2013 Photo: João Vitaliano. Jan/2014

It can be seen in Figure 7 that the volume of sediment accumulated in front of the restaurants prevented the seas from approaching with an amplitude below 2.3. Figure 8 gives an idea of the extent to which these houses were built in the area that should have been the beach. The 3.0 sea level shows not only the removal of sediment from this area, but also the mistake made by urbanisation without prior study of the relationship between the sea and the mainland.

Figure 7 Reach of Mare 2.3/jan2013 Figure 8 Reach of Mare 3.0/jan2014
Photo: Joao Vitaliano. Jan/2013Photo : Joao Vitaliano. Jan/2014

Souza (1980) points out that one of the main causes of coastal erosion is the construction of houses and buildings in the area that corresponds to the beach environment. What was observed in the buildings in Figures 09, 10, 11, 12, 13 and 14 is that the houses were built in the area considered to be post-beach, an area that should be formed by dunes and vegetation cover to receive the first impact of the waves in the high seas that have been responsible for the disfigurement of several beaches on the Brazilian coast.

Figure 09 Restaurante do Francês 1 Figure 10 Post-beach invasion 2
Photo: João Vitaliano.Jan/2014 Photo: João Vitaliano.Jan/2014

Figure 11 Invasion on post-beach 3 Figure 12 Invasion on post-beach 4
Photo: João Vitaliano.Jan/

Figure 13 Invasion of the beach environment 5 Figure 14 Invasion of the beach environment 6

Photo: João Vitaliano. Jan/2014 Photo: João Vitaliano. Jan/2014

Figure 15 shows the phenomenon described by Morais (1980) in relation to engineering works, in this case the spikes which trapped a large amount of sediment, preventing the waves from reaching far enough, but transferred their force to the region downstream, causing a major impact on the houses, as can be seen in Figure 13, located approximately 200 metres after the last structure. Figure 16 shows the sediment completely covering the rockfill, which in 2013 was completely exposed to protect the partially destroyed houses and street.

Figure 15 Spike at 2.7 tide Figure 16 Sediment on the rocks
Photo: João Vitaliano.Jan/2014 Photo: João Vitaliano.Jan/2014

Figure 17 closes the table of images reflecting the current reality of Caponga beach, located in the municipality of Cascavel-CE. The image was taken 150 metres away from the last house built on the beach. It shows the presence of dunes with little vegetation, but still out of reach of the high seas that occurred in January and will recur throughout 2014. The only building we can see in this area is behind the pile of sand to the right of the image, a place not affected by the high seas.

Figure 17 Total Absence of Coastal Erosion in the North Zone of Caponga

Photo: Joao Vitaliano. Jan/2014

The suborbital image shows the group of buildings that have suffered most from the rough seas of recent years on Caponga beach. The image shows that the houses were built on the edge of the beach and have been exposed in recent years due to the changes in the sea, especially in relation to the height of the waves. After the construction of the wickets, which benefited the area located further to the south, on the left edge of the image, the wave energy was diverted to the places indicated by the arrows, causing a great deal of destruction, with large houses completely collapsing and becoming totally unusable. The area in Figure 17 corresponds to the image to the right of Figure 18, where it is possible to compare with the buildings constructed in the post-beach area, the places most affected by

the 3.0 high seas that caused great destruction. In the houses built on the seafront, stones were placed to make up the rockfill, as happened in the Frances restaurant.
Figure 18 shows a suborbital image taken by Google Earth of the area most punished by the waves at high tide. Failure to observe the minimum distance of 150 metres from the post-beach culminated in the disaster seen in many houses on Caponga Beach. It is possible to see the cut in the land made by the real estate company, which does not respect the 150 metre distance indicated by the law. Arrows 1 and 2 show the buildings on the beachfront. Arrow 3 shows how a restaurant has been planned on the beachfront.

Figure 18 Buildings in the post-beach area
Photo: Google Earth. Nov/2012

The arrow in the centre of Figure 19 shows the distance of 70 metres from the houses to the edge of the beach, built in an area protected by reefs, where erosion does not occur, even in the 3.0 seas seen in recent years. The arrow on the right points to the area of the restaurants and neighbouring houses that have had the sediment removed.
in the photo completely shaved in 2013.

Figure 19 Residences in the South Zone and Restaurant Area
Photo: Google Earth.Nov/2012

4 FINAL CONSIDERATIONS

Based on the images presented in this research, it was concluded that it was not the sea's advance over the mainland that was responsible for the destruction of the houses, but the invasion of buildings in the post-beach areas belonging to the sea. From the photo taken by remote sensing, it is possible to see the lack of urban planning by the city council, which has allowed sequences of invasions and constructions in areas considered forbidden, in this case, the post-beach area, not keeping the 150 metres distance allowed by law. Stones are placed to contain the waves that hit the most exposed houses, but it is known that this is not enough to prevent the action of the sea at the base of the houses, which is becoming more and more compromised with each passing year. As we don't know what nature will bring in the coming decades, it's important that the owners of these homes build their houses in more remote areas, thus avoiding occurrences such as those shown in Figure 13.

5 REFERENCES

Ministry of the Environment (MMA). Coastal and Air Quality Management. Summary document of the 1st National Symposium on Coastal Erosion. 24p. Brasilia. 2008.

SOUZA, C R G. Coastal Drift Cells and Erosion on the Beaches of Sao Paulo. Doctoral thesis. 2 volumes. Institute of Geosciences. USP. 1997.

SOUZA, C R G. SUGUIO, K. The Coastal Erosion Risk Zoning and the State Sao Paulo Plan for Coastal Management. Journal of Coastal Research. Special Issue. 35. pp530 - 547. 2003.

MAIA, L.P. Processo costeros y balance sedimentario ao lo largo de Fortaleza (NE-BRASIL): implicaciones para uma gestion adecuada de la zona litoral,1999. Thesis (Doctorate). University of Barcelona-Spain, 269 p. 1998.

MENDES, J N V. PINHO, J L S. Coastal erosion - Methodologies for its quantification. ***Civil Engineering***. N° 33. 2008.

MORAIS, J. O. Aspectos da Geologia Ambiental Costeira no Munidpio de Fortaleza - Ceara - Thesis - 318 pp. Fortaleza, 1980.

CHAPTER 3

ENVIRONMENTAL COMPROMISE IN URBANISED CITIES: STRUCTURAL CHANGES AT BOA VIAGEM BEACH - RECIFE (PE) CAUSED BY ANTHROPIC ACTION

Boa Viagem beach, located in Recife, Pernambuco, is one of the best known urban beaches in Brazil. However, like many urbanised coastal areas, it has faced serious environmental problems as a result of urban development and structural changes over time. Below are some of the main issues related to environmental compromise on Boa Viagem beach:

1 Coastal erosion

Coastal erosion is one of the main problems facing Boa Viagem beach. Infrastructure works, such as the construction of avenues and buildings close to the coast, have altered the natural dynamics of sea currents and sediment transport, accelerating the erosion process. In addition, the construction of retaining walls and other barriers in an attempt to protect the shore ended up aggravating the situation in some stretches, displacing the problem to other areas of the beach.

2 Reduction of the Sand Strip

Structural changes on the Boa Viagem waterfront, including intensive urbanisation and the construction of large buildings, have contributed to the reduction of the beach's sand strip. The reduction of the sand strip not only decreases the space available for recreational activities, but also increases the area's vulnerability to flooding and the action of high seas, putting coastal infrastructure and property at risk.

3 Water and Soil Pollution

With urbanisation, there has been an increase in the pollution of seawater and soil in the region. Domestic and industrial sewage, often inadequately treated, is dumped directly into the ocean, affecting the quality of the water and the health of bathers. In addition, rubbish accumulated on the beach, including plastic and other solid waste, pollutes the marine environment and can be harmful to aquatic life.

4 Heat Islands

Urban densification in the area around Boa Viagem beach, characterised by the construction of large concrete buildings and the reduction of green areas, contributes to the formation of urban heat islands. This phenomenon raises temperatures in the region, making the environment less comfortable for residents and visitors and increasing the demand for energy for cooling, which in turn contributes to greenhouse gas emissions.

5 Impact on Marine Biodiversity

Urbanisation and structural changes on Boa Viagem beach also have a negative impact on marine biodiversity. The construction of barriers and the modification of the coastal environment alter natural habitats, harming species that depend on reef areas and beach zones for feeding and reproduction. In addition, pollution from sewage and solid waste can affect the health of marine ecosystems, reducing the diversity of species.

6 Flood Risks and the Effects of Climate Change

Boa Viagem beach is vulnerable to the impacts of climate change, such as rising sea levels and an increase in the frequency and intensity of storms. Urbanisation and the construction of rigid infrastructure along the coast can exacerbate these risks, making the region more prone to flooding. These events can cause significant damage to urban infrastructure and threaten the safety of residents and visitors.

7 Tourist Pressure and Environmental Degradation

Boa Viagem is a popular tourist destination, which puts additional pressure on the area's natural resources. Mass tourism can contribute to environmental degradation if it is not managed sustainably. Beach overcrowding, increased rubbish generation and pollution caused by uncontrolled tourism can exacerbate existing environmental problems. Structural changes on Boa Viagem beach, driven by urbanisation and economic development, have resulted in significant environmental damage. Coastal erosion, pollution, the reduction of the sand strip, and impacts on biodiversity and quality of life are issues that require urgent attention. To mitigate these problems, it is essential to implement sustainable coastal management policies that include restoring the sand strip, controlling pollution, preserving biodiversity and adapting to climate change. The dune environment with its vegetation in the coastal region has become one of the areas of the geographical space that suffers most from poorly planned anthropic action, leaving it exposed to degradation of its natural components. The preservation of frontal dune ridges is still seen as the best solution for preventing coastal erosion. The beach arch of Boa Viagem Beach can be described as a system dominated by waves and winds, agents responsible for sediment dynamics and modellers of the coastline. Beaches form a dynamic environment in which the elements of nature, combined with the actions of man, build and destroy in a constant process of remodelling the landscape. Some of the alterations imposed on the beach environment by anthropic action over four decades have been recorded. The aim of this study was to assess anthropogenic action in the process of devastating the dunes and vegetation cover of Boa Viagem beach, located in the southern part of the municipality of Recife, Pernambuco.

1 INTRODUCTION

The coastline is considered to be one of the most dynamic features on the planet and its position can be affected by factors of natural origin and intrinsically related to coastal dynamics, such as: sediment swings, variations in the relative sea level, sediment dispersion, and others of anthropogenic origin, such as engineering works, river damming, destruction of sandbank vegetation, dredging, etc. Coastal dunes are eolian landforms that develop in situations where a large supply of loose, sand-sized sediment is available to be transported inland by the winds of the environment (Martinez et al., 2004). Dunes serve as protection to prevent the progression of waves and vegetation is an important support for their existence (Hesp & Walker, 2012).

The dune environment on the coastline has become one of the parts of the geographical space that suffers most from poorly planned anthropogenic action, leaving it exposed to degradation of its natural components (Sousa et al. 2012). Beaches are a dynamic environment and the available forecasts suggest a wide range of impacts, from changes in morphodynamics due to anthropogenic action, incident energy, changes in sediment composition, slope and area available for occupation by organisms, in milder and/or initial conditions, to the loss of the intertidal and post-beach region and, consequently, its biota associated with extreme and recurrent events (Rosa Filho, 2015).

Frontal dunes are formed by the deposition of sediments carried by wind action in the post-beach vegetation area (Hesp, 1989); (Hesp et al., 2005). Dunes are very sensitive to the environment in which they are formed because they are composed of light sediments that are easily carried away by wind action (Lourengo, 2012). Dunes are hills of sand that vary in height and are formed by the action of the wind, which have areas covered or not by vegetation that prevent the action and progression of the sea by fixing the sediments brought by wind action and can be separated from each other by a few metres (Ubeid & Albatta, 2015).

This paper refers to Boa Viagem beach, which is located in the southern part of the Boa Viagem neighbourhood and is considered the most frequented urban beach in the city of Recife. According to IBGE data (2000), its geographical density is 136.0 inhabitants/ha and the increase in its population was due to the replacement of houses by buildings with more than 20 floors from the 1970s onwards (ARAUJO, 2008, p.53). Throughout the coastal environment, the high number of these buildings filled the spaces that would have been used for public green areas, as well as allowing natural coastal processes such as strong winds and the reach of waves in sizzling seas to occur uninterruptedly (Araujo, 2008, p.62). The northern part of Boa Viagem beach has herbaceous cover, forming an environment susceptible to anthropogenic impacts, mainly due to the constant trampling caused by beach users who cross the vegetation forming paths that encourage wear and tear (Araujo, 2008, p.76).

The techniques of visual image interpretation are widely used to record landscapes and the changes observed in the environment over the years. These are homogeneous delimited regions that will form the basis of the work of visual interpretation of old and current aerial photos, local photos and satellite images (Panizza & Fonseca, 2011). Coelho et al. (2000); Rocha et al. (2007); Sa (2002) state that the restinga and mangrove vegetation used to make up the native vegetation of the beaches, which anthropogenic action has altered, giving way to coconut plantations, a type of crop that is adaptable to the region due to the sandy soil found in the submerged areas and the large stretches of shallows under the influence of the seas.

Costa et al. (2008) define an urban beach as a set of structures made up of pavements, play areas for children, courts, football pitches, gardens and buildings that become part of the landscape as part of the social nature that directly or indirectly affects natural ecosystems. Intense verticalisation in the Boa Viagem neighbourhood has brought irreversible environmental consequences, such as: the construction of spiky buildings, the formation of heat islands, increased sewage flow, waterproofing of the neighbourhood, disfigurement of the landscape, uncontrolled traffic, urban atmospheric pollution by particulate matter, increased demand for water resources, impeded wind circulation and shade on the beach during the afternoon. According to the same author, the first building constructed on the beach was in 1958, but by 2007, 201 had already been registered, 85% of which were over 10 storeys high (Costa et al., 2008).

The coastal vegetation of Boa Viagem beach, which extends as far as Pina beach, was considered a Special Environmental Protection Zone by Recife City Hall, but the construction of the avenue, promenade, courts and leisure areas influenced the sealing of the land and, consequently, the mobilisation of the dunes, influencing the displacement of the sedimentary balance (Costa et al., 2008). Figure 1 represents the model of frontal dune formation and remodelling described by Hesp (2002), which includes the process of degradation and recomposition of the frontal dune cordon that has been altered by anthropogenic or natural phenomena. A process of coastal zone formation that took thousands of years to be structured was completely destroyed by anthropogenic action, starting in the 1980s with the Cura Beira-Mar project under the responsibility of the Recife Urbanisation Company (URB), whose aim was to restructure the waterfront in an attempt to offer better accommodation for tourists and the population. However, the interventions caused wear and tear on the beach environment as a result of the devastation of the frontal dunes with the vegetation cover, lowering the topography of the beach strip and potentiating the natural erosion process (SILVA, 2016, p.73).

Sediment flow structure in natural conditions of a dune by Hesp (1981); Hesp (2002):

1. The shape of the vegetation;
2. The height of the vegetation;
3. The density of vegetation;
4. The distribution of vegetation;
5. Wind speed;

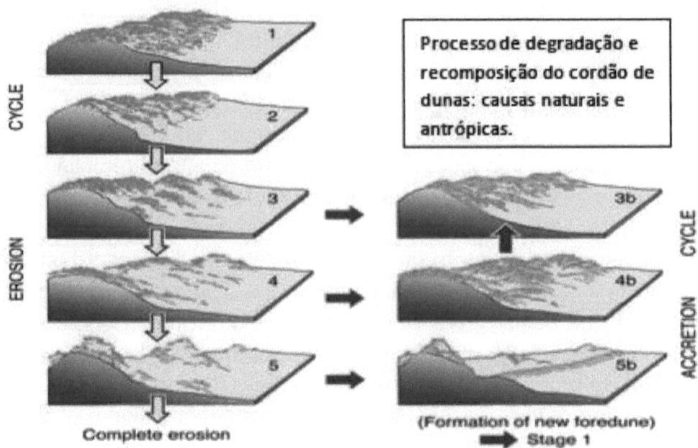

The process of dune cordon degradation and recomposition: natural and anthropogenic causes.

Figure 1: Moments observed in the erosion and recomposition process of the dune system. Source: Hesp (2002).

Hesp (1988) mentions that dunes follow a medium to long-term evolutionary model divided into five stages, which can be applied to progressing, stable or eroding coastlines. The stages comprise: Stage 1 is characterised by dunes with simple topography, laterally continuous and well vegetated (90-100%); Stage 2 is characterised by less vegetation cover (75-90%), lateral variation in vegetation density (alternation of places with and without vegetation, forming channels or openings along the cordon) and the presence of erosional scarps at the base of the dunes; stage 3, which is the most common in coastal frontal dunes (HESP, 1988), has a *hummock-type* topography characterised by a fragmented, irregular, asymmetrical morphology, with monticular forms and less vegetation cover (45-75%), which favours the formation of small to medium blowouts; stage 4 is characterised by partially vegetated ridges (20-45%) and greater topographic variability than stage 3, with medium to large blowouts as well as sand plains and deflation basins; and stage 5, where a large part of the dunes have been removed by wind and/or wave action, with only remnant mounds and segments of the ridges remaining.

1.1 Coastal Vegetation Cover - Permanent Preservation Area

Article 3 of Law No. 7.803 of 18 July 1989 deals with Permanent Protection Areas, which include forests and other forms of natural vegetation declared by the government to mitigate land erosion and fix dunes (Machado, 1991). The anthropogenic impact caused by misuse of vegetated areas can lead to modifications or destruction of the ecosystem. High levels of disturbance result in a decrease in vegetation height and cover, leaving the land exposed and vulnerable to erosion (Hesp, 2002). Despite the impacts that have occurred in this ecosystem, few studies have been carried out on the destruction of dunes with vegetation cover and, above all, on the restoration of destroyed areas (Rosa & Cordazzo, 2007). The aim of this study is to carry out a survey of anthropogenic action on the dunes and vegetation cover of Boa Viagem beach (PE) over the last three decades. The vegetation cover of the dunes is the main natural element capable of efficiently protecting the soil and the relief, playing an important role in the formation and fixation of the dunes, and its variation depends on the degree of tolerance of the plant species in relation to the deposition of sand on them.

Leite & Andrade (2004) found 47 plant species covering the dunes distributed in 20 families and 43 genera, the vast majority of which were herbaceous, among them the species Crotalaria retusa (purple loosestrife), Ipomoea littoralis (white beach parsley), Ipomoea pes-caprae (purple beach parsley), Turnera ulmifolia (chanana), Sporobulus virginicus and

Paspalum maritimum are considered to be dune fixers. In analyses of herbarium collections, Leite & Andrade (2004) found that the species R. maritima was present on Boa Viagem beach until 1987. The vegetation cover that makes up the dune structure along Boa Viagem beach is considered, according to the current Forestry Code, Law 12.651/12, to be a permanent preservation area, APP, because it fixes the dunes and mitigates soil erosion (Lourengo, 2012).

A permanent preservation area can be defined as a protected area, whether or not covered by native vegetation, with the environmental function of preserving water resources, the landscape, geological stability and biodiversity, facilitating the genetic flow of fauna and flora, protecting the soil and ensuring the well-being of human populations (Ministry of the Environment, 2013).

2 METHODOLOGY

The purpose of this study is to present the degradation of the frontal dunes that has occurred since 1989 on Boa Viagem beach, with the knowledge and approval of the municipality's public administration. To carry out this study, the Google site was used as a search engine, from which articles from important theoretical contributions were extracted, such as: Patrick Hesp, Rosa & Cordazzo, Martinez, Araujo, among others cited in the references. The field trips to obtain the images took place in 2011 between the months of July and September.

2.1 Characterisation of the study area

Boa Viagem beach is located in the Boa Viagem neighbourhood and has always been used as a summer resort with a few secondary houses. It is bordered to the north by the BrasHia Teimosa neighbourhood and the Port of Recife, to the south by Piedade beach, which belongs to the municipality of Jaboatao dos Guararapes, to the west by the mangroves and to the east by the Atlantic Ocean (Silva, 2000). Figure 2 shows the area of the municipality of Recife, in which the area of the Boa Viagem neighbourhood, in which the beach that is the subject of this study is delimited, is shown in red in the smaller box on the right. The image in Figure 2, which corresponds to the length of Boa Viagem beach, shows the removal of 1,850.33 metres of frontal dunes that had been impacting sediment for decades.

Figure 2 - Boa Viagem neighbourhood and beach, south side of the city of Recife/Pernambuco

, Brazil.

Source: https://mapasblog.blogspot.com/2012/03/mapas-de-recife-pe.html.

Despite knowing that some interventions to recompose the frontal dunes would not be possible because they do not represent works of great importance for the propaganda of the municipality's administration, the vulnerability assessment of the areas whose resilience has been threatened in recent decades was based on the concepts proposed by Tabajara, Gruber, & Portz (2012) using parameters that summarise the systematic condition of the dunes, including: surface area, scarping process, evolutionary stages and slope on the marine front. The main parameters that summarise the systemic condition of the dunes were given a score from 0 to 4, with the first representing an equilibrium condition and the last an advanced stage of degradation. The parameters were grouped into four sections:

Section A (dune morphology) refers to the possibility of dunes regenerating after environmental stress events and is related to their size and volume, so the larger the system and the greater the availability of sediment, the better the ecological and geomorphological relationships will be, along with protection against wind and sea erosion;

Section B (beach condition) refers to the post-beach and is related to the supply of sediment to the frontal dunes. The width of the beach is directly proportional to the availability of sediment.

Section C (characteristic of the 200 m adjacent to the sea) contains more comprehensive parameters that synthesise the bioffsic environment. Blowout areas within the system, gaps in the dune system;

Section D (use pressure) corresponds to anthropogenic elements that directly affect the balance of the frontal dunes. Presence of a kiosk on the beach, urbanisation furniture, vehicular access, pedestrian paths.

Field images were taken using a Sansung Lens ES95 camera and Google Earth images to show the extent of the destruction of the dunes with vegetation cover. To obtain the data, the Google website was used as a search engine. Due to the lack of adequate information to characterise the observations presented in this research, we used the Google Earth image in which we used the ruler available in the table shown in Figure 3 to establish the extent of the complete destruction of the dunes with vegetation cover in the central area of Boa Viagem beach in the 1980s and which remains to this day. The yellow marker on the right shows the boundary between the destroyed and preserved vegetation, located thirty metres from Rua Prof. Jose Brandao, and the marker on the left is set in the area next to the old Castelinho, which has been transformed into a condominium with two towers over twenty storeys high. The total length of the yellow line reaches 1,850.33 metres, an area of dunes less than one metre high and devastated vegetation cover with no possibility of natural recovery in the short term or in the interest of the bodies responsible for preservation.

The marker used by Google Earth shows the entire length of Boa Viagem beach without the frontal dunes, which were removed after a project at the end of the 1980s. Coastal erosion began in 1994 and continued until 2011. The scraping of the sediments became intense, which caused a significant drop in the vicinity of D. Lindu Park, extending to the immediate vicinity of the Castelinho residential complex.

Figure 3 Extent of Destruction of 1,850.33m of Dunes with Vegetation Cover in Boa Viagem.

3 RESULTS

Figure 4 shows the features of Boa Viagem beach in three different decades. In the 1970s, the first photo shows the integrity of the dunes with vegetation cover. The images from the 1990s and 2000s clearly show the scraping of the vegetation and the dunes, registering environmental degradation. To date, no one has taken responsibility for this ecological disaster. It is possible to observe the rapid verticalisation of this neighbourhood in 40 years, with buildings replacing houses. The dunes, ranging in height from 80 cm to 1.20 metres, were covered by vegetation made up mostly of beach grass and parsley. The removal of this important component of the landscape favoured the displacement of sediment from the south to the north, as well as its projection onto the beach and avenue, especially in August when the winds are considered strongest. Figure 4 shows that the strip of land in the first image was covered by vegetation until 1989, when the project that culminated in its complete destruction was completed. The second and third images, taken from 1990 and 2000 respectively, show no recovery of the vegetation or recomposition of the frontal dunes. In August 1994, the

coastal erosion on the south side of Boa Viagem beach.

Environmental degradation of the Boa Viagem beachfront in the 1970s, 1990s and 2000s.

Source: Adapted from Silva (2016)

The mangrove area, once rich in fish and crustaceans, at the top of the photo in Figure 5 now receives much of the contaminated waste from these habitats, giving the environment a nauseating odour. The life forms found before, in this very important ecosystem, no longer exist.

Figure 5 - Panoramic view of the dunes with vegetation on Boa Viagem beach in the 1970s.

Source: https://divulgapernambuco.webnode.com.br/galeria-de-fotos/imagens-antigas/

The images in Figures 6 and 7 are inserted in the stretch of land shown in the three images in Figure 4. All the vegetation has been decomposed to make way for volleyball courts and space for bathers, in large numbers, to lounge. Nothing has yet been done to restore these degraded areas. Figure 6 shows how abundant the vegetation and dunes were until the 1980s. At the top of Figure 7, the blue arrow shows that there was no erosion until the end of the 1980s. The white arrow shows the first traces of environmental degradation of the frontal dunes.

Figure 6. Vegetation cover and dunes in front of the Acaiaca building.

Source:https://www.youtube.com/watch? v=pTqcLL-GW8A

Figure 7. Destruction of frontal dunes.

Source:https://divulgapernambuco.webnode.co m.br/galeria-de-fotos/imagens-antigas/.

Figures 8, 9 and 10, located on the north side of Boa Viagem beach, show three different areas in which the traffic of people is intense every day of the week, preventing vegetation from growing.

Figure 8. Frontal dunes Figure 9. Pisoteio 10. Destruction Photos: Research data by João Vitaliano Nov/2013 Figure 8 shows on the north side of the beach the frontal dunes with vegetation that have been preserved and which, have never been hit by the spring tides, even on the tides of 2011 which reached heights of 2.6 and 2.7 considered abnormal for this region, in several months. It can be seen in these tides that the bare sand is easily swept away because the dunes don't have the vegetation to impact it, being pushed in the direction of the cliff, essentially by wind action, and can invade and even cover land, projecting sand onto pavements and avenues, houses and paths (Sousa et al. 2012). Figure 9, called trampling, shows a very extensive area of destruction of the dunes, with the native vegetation coming from the crowds that visit the beach every day, which is more intense on Saturdays, Sundays and public holidays. Figure 10 shows a cut in the dunes with vegetation for more than 40 metres, and it is possible to make out the marks of car tyres that are taken to the post-beach area. Between the months of August and September, the winds blow more intensely, blowing the sediment onto the beach and avenue, as shown in Figures 11 and 12. The groove formed next to the protection wall is made by the city council in an attempt to stop the sediment moving easily from the post-beach in the absence of dunes and vegetation. Recife City Hall digs a kind of trench every year from July onwards, when the strongest winds begin, peaking in August and gradually diminishing in September, to receive the sediment brought in by the winds. The moments seen in Figures 11 and 12 correspond to the action of the winds in July and it is possible to see the deposit of sediment in the marsh that extends to the avenue. Recife's City Hall has been called upon to minimise the impact of this disturbance. This phenomenon only began to occur after the dunes were removed.

Figure 11. Area without dunes. Figure 12. Sediment impediment.
Photos: Joao Vitaliano. August/2011.

3.1 Consequences of Recreational Activities and Pedestrianisation

Recreational activities can cause accumulative erosion, which advances on the adjacent vegetation, reducing its density, as well as affecting its capacity to accumulate sand on the vegetation, resulting in a de-characterisation of the entire environment, which ends up further increasing the erosive potential, leading to a large area with loss of vegetation cover (Rosa & Cordazzo, 2007). Occupation almost always occurs in areas of coastal and restinga vegetation, characteristic of Atlantic forest vegetation, as well as sandbanks (Mendes & Barcellos, 2018).

Anthropic action in the dune area can also be seen in the trails formed by trampling by bathers, which are part of a series of geological evolutionary processes that reflect on coastal morphology. Constant trampling leads to soil compaction, increasing its resistance to penetration, according to Rosa & Cordazzo (2007). According to Talora (2007), the impact of trampling can cause the exclusion of native species and the possibility of the emergence of invasive species that could alter the population dynamics of vegetation over large areas. Figures 13, 14, 15, 16, 17 and 18 represent the sequence of the formation of more extensive openings between the dunes and the vegetation cover that end in areas of great destruction due to constant trampling. Figure 13 shows the beginning of the destruction in an area with native vegetation cover, where people walk freely in search of other accesses to the beach. Figure 14 shows the increased wear and tear on the vegetation that occurs when thousands of people use this path for months or years without worrying about the wear and tear. Figure 15 shows the lack of environmental awareness when soft drink sellers open up a dune without permission and often set fire to the undergrowth. Figure 16 shows areas of wear and tear on the vegetation and a new path opened up in the frontal dunes. The image in Figure 17 shows another opening in the dunes that is widely used by carts and people to access the edge of the beach. Figure 18 shows the harsh reality of the constant trampling that leads to the formation of the Blow-out, an area that has been completely destroyed despite being considered a "Permanent Preservation Area".

Figure 13. Path through vegetation. Figure 14 Expansion of the path. Figure 15. Opening in the Dunes
Photos: João Vitaliano. Nov/2013

Figure 16. Start of Destruction Figure 17. Wagon track Figure 18. Blow-Out
Photos: Joao Vitaliano. Nov/2013 (Survey data)

4 DISCUSSION

Coastal environments, although dynamic, are extremely fragile and have been used in recent decades for recreation, leisure and tourism, activities that accelerate the deformation of the landscape, mainly due to trampling and the recovery of frontal dunes (Macedo, 2001). The cultural valorisation of these areas by the real estate market because of their landscapes has favoured the proliferation of vertical development, condominiums, hotel structures and the unbridled urbanisation of the most vulnerable sectors of the coastal environment (Strohaecher, 2007). Portz (2010) asserts that the removal of dune vegetation results in the loss of sand due to wind and sea action, making the region more vulnerable and exposed to stress, making it difficult to recover (Adger, 2006). The position of urbanisation, which occupies the post-dune area, is associated with a high density of pedestrian streets, which degrades the vegetation and releases sand for wind transport to the pavements, avenues and streets due to the lack of pioneer vegetation (Hesp, 2000). Waves also have a major impact on dune erosion (Larson, Erikson & Hanson, 2004); (Overton, Fisher & Young, 1988); (Palmsten & Holman, 2012). The oscillations in the movement of sediments are progressive in the direction of the land (Sallenger & Richmond, 1984), and with the removal of coastal vegetation the winds delineate the coastline as described (Mahiques et al., 2010); (Campos & Dominguez, 2010); (Luz, Barth & Silva, 2005), mainly because they do not encounter natural obstacles (Antony, Vanhee & Ruz, 2007). The year 1989 was a landmark for Boa Viagem beach, which suffered destructive action from some beachgoers who asked the tractor operator responsible for cleaning the area to remove a vast area of beach grass, native vegetation that is responsible for the impact of sediment projected by the wind from the post-beach. In the absence of vegetation, there is a phenomenon, as described by Sousa (2012), of bare sand that is easily dragged inland, either by the action of the sea or, essentially, by wind action, and can invade and even cover agricultural land, projecting sand onto pavements and avenues, houses and roads. Figure 5 shows one of the worst affected areas of Boa Viagem beach, where coconut palms have been planted almost the entire length to provide shade for bathers and visitors. Figure 6 shows the scraping of an area to build beach volleyball courts, a trend that continues to this day. Figure 8 shows the vegetation in the post-beach area that was not affected. Figure 9 shows an area of trampling in which an artesian well was drilled to attract bathers, making it a busy place where no more vegetation grows. Figure 10 shows the destruction of the vegetation, causing an opening and breaking the entire segment of an area that was covered in vegetation until the early 90s. Figures 11 and 12 show the consequence of the removal of the dunes with the native vegetation, causing the sand to slide into the limestone. Figures 12 to 17 characterise the negligence of the environmental authorities, coupled with the ignorance of beach-goers of the need to conserve a natural environment. Destruction often begins with the formation of a simple path through the vegetation. Then vendors drive through with their carts, widening the area that serves as a passageway to the edge of the beach. The purposeful openings culminate in the formation of blow-outs, which characterise the accentuated degradation of native vegetation.

5 FINAL CONSIDERATIONS

In just four decades, the structure of the beach environment has been almost completely destroyed by the lack of environmental education among its visitors and by the competent authorities responsible for its maintenance. The images presented here were taken in December 2013 and show the stages of wear and tear to which the dunes are subjected. The coastal vegetation that protected the area was completely devastated by the end of 1989, favouring the transport of sediment from the south to the north, further compromising the coastal erosion that began five years after the dunes were removed. The dunes with their native vegetation to protect against wind erosion were partly destroyed without any prior study of the consequences for the natural environment. The adoption of an occupation model without an adequate urbanisation project has facilitated the invasion of the property market and the implementation of a project to clear the vegetation from the beach arch by the government has favoured a significant loss of dunes with vegetation cover and that there is no adequate management for the areas that are being eroded every year, as well as interest in the implementation of recovery projects by local universities and the bodies that make up the town hall. The aim of this study was to understand the consequences of anthropogenic action in a natural environment in which the dunes with vegetation cover on Boa Viagem beach in the municipality of Recife have been partially destroyed.

6 REFERENCES

Antony, E. J.,Vanhee, S.,&Ruz, M.H. (2007). An assessment of the impact of experimental brushwood fences on foredune sand accumulation based on digital elevation models. *Ecological engineering, 3*(1), 41 -46.https://doi.org/10.1016/j.ecoleng.2007.05.005

Araujo, M.C.B. (2008). *Boa Viagem Beach, Recife-PE: socio-environmental analysis and planning proposals* (Doctoral thesis, Federal University of Pernambuco, Recife, Brazil). Retrieved from https://repositorio.ufpe.br/handle/123456789/8058

Ministry of the Environment (MMA) (2013). Forestry Code. *Interinstitutional Working Group. Restoration and preservation of permanent preservation areas (APP).* Recovered fromhttp://www.mma.gov.br/port/conama/processos/E72A2846/DispositivosLegaisAPP.pdf

Campos, R.H.S. & Dominguez, J.M.L. (2010). Mobilityofsedimentsduetowaveactiononthe continental shelfoftheNortherncoastofthestateof Bahia. *Brazilianjournalofoceanography, 58* (spe2), 57-63.http://dx.doi.org/10.1590/S1679-87592010000600007

Cerqueira, R. Biogeography of the Restingas. In: Esteves, F A. Lacerda, L D. (eds.) Ecologia de Restingas e Lagoas Costeiras, Rio de Janeiro: Macae. NUPEM/URRJ, 2000.

Coelho, G. Valcarcel, R. Neves, L G. Pin, B D. Project for the recovery of degraded areas and environmental control plan for a campsite in an APA on Ilha Grande, Angra dos Reis. In: Simp. National Symposium on the Recovery of Degraded Areas, IV Abstracts..., 285p. p.208-209. SOBRADE. Blumenau, SC.

Costa, M. F.da, Araujo, M.C.B., Silva-Cavalcanti, J.S., & de Souza, S.T. (2008). Verticalisation of Boa Viagem Beach (Recife-Pernambuco) and its Socio-Environmental Consequences. Revista da gestao costeira integrada, *8*(2), 233-245.https://doi.org/10.5894/rgci128

Hesp, P A. Coastal sand Dunes - Form and Functions. CDVN Technical Bulletin. N°4, Palmerston North, New Zealand: Rotorua, 2000. In: Gravel, n°2. Porto Alegre, CECO/IG/UFRGS, p.104-121. 2004.

Hesp, P. (2002). Foredunes and blowouts: initiation, geomorphology and dynamics. *Geomorphology, 48*(1-3), 245-268.https://doi.org/10.1016/s0169-555x(02)00184-8

Hesp, P.A., Dillenburg, S.R.,Barboza, E.G., Tomazelli, L.J., Ayup-Zouain, R.N., Esteves, L.S., ...Clerot, L.C.P.(2005). Beach ridges, foredunes or transgressive dunefields? Definitions and an examination of the Torres to Tramanda^ barrier system, southern Brazil. *Anais da academia brasileira de ciencias, 77*(3), 493-508.http://dx.doi.org/10.1590/S0001-37652005000300010

Hesp, P.A., & Walker I.J.(2012). Three-dimensional aeolian dynamics within a bowl blowout during offshore winds: Greenwich Dunes, Prince Edward Island, Canada. *Aeolianresearch, 3*(4), 389-399. https://doi.org/10.1016/j.aeolia.2011.09.002

Hesp, Patrick A. The formation of shadow dunes. Journal Sedimentary research. 1981. 51(1):101-112. https://doi.org/10.1306/212F7C1 B-2B24-11D7-8648000102C1865D

Hesp, Patrick A. A review of biological and geomorphological processes involved in the initiation and development of incipient foredunes. Volume 96. 1989, pp.181 -201. https://doi.org/10.1017/S0269727000010927

Hesp, Patrick A. Morphology, dynamics and internal stratification of some established foredunes in southwest Australia. Sedimentary geology. Vol.55, Issue 1-2, 10, march, 1988, pp.17-41. https://doi.org/10.1016/0037-0738(88)90088-7

Larson, M., Erikson, L., & Hanson, H. (2004). An analytical model to predict dune erosion due to wave impact. *Coastalengineering, 51* (8-9), 675-696.https://doi.org/10.1016/j.coastaleng.2004.07.003

Lourengo, C.S. (2012). *Dune front dynamics as a function of human interventions* (Master's thesis, University of Aveiro, Aveiro, Portugal) Retrieved from https://ria.ua.p1/bitstream/10773/9808/1/disserta%C3%A7%C3%A3o.pdf

Luz, C.F.P. da,Barth, O.M.,&Silva, C.G. (2005). Spatialdistributionofpalynomorphs in thesurfacesedimentsofthe Lagoa do Campelo lake, North regionof Rio de Janeiro State, Brazil. *Acta BotanicaBrasilica, 19*(4), 741-752. http://dx.doi.org/10.1590/S0102-33062005000400009

Macedo, SHvio Soares. Production of the contemporary urban landscape at the end of the 20th century. Landscape and Environment. 143-170. 2001.

Machado, P.A.L. (1991). *Brazilian Environmental Law*. Sao Paulo: Revista do Tribunais Ltda.

Mahiques, M.M. de, Sousa, S.H. de M. e, Furtado,V.V., Tessler,M.G., Toledo, F.A. de L., Burone,L., ... Alves, D.P.V. (2010). The Southern Brazilianshelf: general characteristics, quaternaryevolutionandsedimentdistribution.*Brazilianjournalofoceanography, 58*(spe2), 2534. http://dx.doi.org/10.1590/S1679-87592010000600004

Mendes, T.M. & Barcellos, C. (2018). The territorial dimension of sanitary sewerage: the case of Recreio dos Bandeirantes, Rio de Janeiro, Brazil. *Ciencia e saude coletiva, 23*(2), 647-658.http://dx.doi.org/10.1590/1413-81232018232.27732015

Overton, M.F., Fisher, J.S., Young, M.A. (1988). Laboratory investigation of dune erosion. *Journal of waterway, harbour, coastal and ocean engineering, 114*(3), 367-373. **https://doi.org/10.1061/(asce)0733-950x(1988)114:3(367)**

Palmsten, M. L. & Holman, R.A. (2012).Laboratory investigation of dune erosion using stereo video.*Coastal engineering, 60*, 123-135. https://doi.org/10.1016/j.coastaleng.2011.09.003

Panizza, Andrea de Castro. Fonseca, Fernanda Padovesi. Visual Image Interpretation Techniques. GEOUSP - Espago e Tempo, Sao Paulo, n°30, pp.30-43, 2011.

Portz, Luana Carla. Manzolli, Rogerio Portantiollo. Gruber, N L S. Correa, I C S. Tourism and degradation in the coastline of Rio Grande do Sul: conflits and managment. Development and Environment, 22, 153-166.2010.

Portz, Luana. Jardim, Joao Pedro de Moura. Manzolli, Rogerio Portantiolo. Gruber, Nelson Sambaqui. Impacts on the Dune System: Natural Dynamics versus Anthropogenic Interference. Ambient.soc. vol.19,n°3, Sao Paulo, July/Sept. 2016.

Rocha, C F D. Bergallo, H G. Van Stys, M. Alves, M A S. Jamel, C E. The remnants of restinga habitats in the Atlantic Forest of the state of Rio de Janeiro, Brazil: habitat loss and risk of disappearance. Braz. J. Biol. Vol.67, n°2. Sao Carlos, May 2007. **http://dx.doi.Org/10.1590/S1519-69842007000200011**

Rosa Filho, J S., et al. Long-term monitoring of benthic microfauna between sandy beach seas. In> Turra, A., and Dendai, Mr., eds. Protocols for monitoring coastal benthic habitats - Coastal Benthic Habitat Monitoring Network - ReBentos[online]. Sao Paulo:Oceanographic Institute of the University of Sao Paulo, 2015, pp.194-208. http://books.scielo.org/id/x49kz/pdf/turra- 9788598729251-15.pdf

Rosa, L.S. & Cordazzo, C.V. (2007). Anthropogenic disturbances in the vegetation of the dunes of Praia do Cassino (RS). *Cadernos de ecologia aquatica, 2*(2): 1-12. Retrieved from http://www.cadernos.ecologia.furg.br/images/artigos/16 Rosa&Cordazzo.pdf

Sa, Cyl Farney Catarino. Regeneration of a stretch of restinga forest in the Jacarepia State Ecological Reserve. Saquarema, State of Rio de Janeiro: II - Shrub layer. Rodriguesia. 53(82):5-23.2002.

Sallenger, A.H. &Richmond, B.M. (1984). High-frequency sediment-level oscillations in the swash zone. *Marine geology, 6*(1-4), 155-164.https://doi.org/10.1016/0025- 3227(84)90148-8

Silva, Cleiton Ferreira. Urban intervention and land use in the southern zone of Recife: an analysis of the urban transformations of the Pina and Boa Viagem neighbourhoods. Adapted from Caderno de geografia, v.26, n°45, 2016. DOI:10.5752/p.2318-2962.2016v26n.45p.55

Strohaecker, T M A. Urbanisation on the North Coast of Rio Grande do Sul: Contribution to Urban Environmental Management in the municipality of Capao da Canoa. 2007. 398p.

Thesis (Doctorate in Sciences) - Postgraduate Programme in Geosciences, Institute of Geosciences, Federal University of Rio Grande do Sul, Porto Alegre, 2007.
Tabajara, L.C.A., Gruber, N.S., & Portz, L.C. (2012). Frontal dunes of Xangri-La, north coast of RS: inventory, classification and choice of priority areas for management. *Geoscience Research, 39*(1). Retrieved from http://seer.ufrgs.br/index.php/PesquisasemGeociencias/article/view/35812
Talora, D.C. (2007). *Effects of experimental trampling on dune vegetation in the Serra do Mar-Picinguaba State Park. Ubatuba.* (Doctoral Thesis, University of Sao Paulo, Sao Paulo, Brazil) Retrieved from http://www.teses.usp.br/teses/disponiveis/11/11150/tde-18042007-143653/en.php
Ubeid, K.F.&Albatta, A.S. (2015). Sand dunes of the GazaStrip (southwestern Palestine): morphology, textural characteristics and associated environmental impacts. *Earth sciences research journal, 18*(2), 131-142. http://dx.doi.org/10.15446/esrj.v18n2.37238

CHAPTER 4

GRAVITATIONAL LANDSLIDES: A CONSEQUENCE OF THE INCREASE IN GLOBAL PRECIPITATION?

Gravitational landslides, also known as mudslides, are movements of soil, rocks and other materials downhill under the influence of gravity. These phenomena can be triggered by various factors, and increased rainfall is one of the most significant.

Consequences of Increased Global Precipitation:

1. **Soil saturation**:

   o As rainfall increases, the soil can saturate quickly. When the soil reaches its maximum absorption capacity, water begins to accumulate on the surface, increasing the weight of the surface layer. This can lead to a weakening of the cohesion between soil particles, facilitating sliding.

2. **Vegetation change**:

   o Vegetation plays a crucial role in soil stabilisation. With global climate change, there are changes in vegetation patterns, such as deforestation and fires, which can expose the soil to greater erosion. The lack of roots to hold the soil facilitates the triggering of landslides.

3. **Increased Frequency of Extreme Events**:

   o Climate change has been associated with an increase in the frequency and intensity of extreme weather events, such as torrential rain. These events can lead to mass landslides in areas where they used to be rare, increasing the risks for people living on slopes or in mountainous areas.

4. **Accelerated Erosion**:

   o Heavy rainfall can cause accelerated soil erosion, especially in areas without adequate vegetation cover. This process can destabilise slopes, making them more prone to landslides.

5. **Risk to Infrastructure and Populations**:

   o In urbanised regions, the increase in landslides due to heavy rainfall can represent a major risk to buildings, roads and other infrastructure. In addition, populations living in risk areas are more vulnerable to disasters, with the potential for loss of life and property.

Events such as the landslides that occur during the monsoon season in south-east Asia or the landslides associated with hurricanes and tropical storms in the Caribbean and Central America illustrate the consequences of increased rainfall on mass landslides.

Mitigation Measures:

- **Monitoring and Early Warning**: Implementation of early warning systems to identify risk areas and predict possible landslides.
- **Reforestation**: Promote reforestation in vulnerable areas to stabilise the soil.
- **Stormwater Management**: Improving drainage systems to reduce soil saturation.
- **Urban planning**: Avoid building in high-risk areas and relocate vulnerable populations.

The increase in global precipitation, associated with climate change, intensifies the risk of landslides, making it essential to adopt mitigation and adaptation strategies to reduce the impacts of this natural phenomenon. Studying the relationship between the increase in the number of mass movement events and the increase in rainfall over the last few decades is the primary objective of this research. Mass movements are important geomorphological processes and are considered to be one of the most elementary processes of relief transformation that occur on the entire surface of the globe. The various forms of mass movement have been presented, as well as anthropogenic action and rainwater as a determining factor in the increase in these events. Several researchers cited in this paper have carried out studies of historical series between 1900 and 2017 in various countries and have concluded that there has been a significant increase in global rainfall, capable of promoting environmental disasters with the highest values for the last two decades.

1 INTRODUCTION

Geomorphology identifies landforms by studying their origin, structure, the nature of the rocks, climate and the endogenous and exogenous factors responsible for shaping or forming certain elements of the Earth's surface. The relief of the Earth's surface is the result of the interaction between the lithosphere, atmosphere, hydrosphere and biosphere, i.e. the processes of energy and material exchange that take place at this interface, in time and space (Trentin et al., 2012).

Among the changes that can occur in a given relief area is gravitational mass movement, which can be defined as the downward and outward descent of slope-forming material under the influence of a transport agent such as water, ice and snow that drags debris flows, terracotta, solubility lobes and rocks (Goudie, 2004).

Popp (2010, p.70) asserts that a large part of the events linked to mass movements occur due to weathering with its hydraulic and chemical reactions that act on the surface materials, weakening the surface rock and making it more susceptible to the force of gravity. In addition to the heavy rainfall, the removal of vegetation destabilises the terrain. The weathering mantle consists of a mixture of rock fragments, sand and clay, organic material and other components (Penteado, 1980).

Mass movements are important geomorphological processes that are considered to be one of the most elementary processes of landform transformation that occur on every surface of the globe and can be particularly pronounced in rugged terrain that is favoured by gravity. With the help of surface water, these movements can occur from small particles to larger fragments, which result in transport processes such as creep, solifluction, landslides and slides, the speed of descent of which depends on the large volume of precipitation and the vegetation cover of the area (Popp, 2010, p.70).

The force that drives the movement of the mass surface is gravity, as it acts in conjunction with the weight, size and shape of the surface material, the degree to which the slope exceeds its angle of repose, the amount of moisture available, whether frozen or fluid, and the degree to which the slope exceeds its angle of repose, the greater the angle of inclination, the more susceptible the surface material is to mass loss processes, however, in some landslides there is the shear force of the slope material, i.e. its cohesion and internal friction act against gravity and mass loss (WICANDER, 2009).

Christopherson (2012) explains that clays, leaflets and mudstones are highly susceptible to hydration, swelling in response to the abundant presence of water. Thus, if such materials underlie rock strata on a slope, the strata will move with less driving force energy.

The term mass or ground movement has been used for all gravitational movements, in which case gravity is the only important force and other types of transport are not involved, such as wind and water, which act as a lubricant to reduce the friction coefficient. Tavena (1971) considers that in any mass movement, gravity exerts a force on the mass until the critical point of failure is reached.

1.1 Geomorphological Science

Geomorphology is a branch of Geography that studies the remodelling of the earth's landforms in order to understand their origin, evolution, shape and spatial distribution (Stetler, 2014). According to Torres et al. (2012), geomorphological science seeks to understand the past and present processes that act on the planet's surface, i.e. the geomorphological analysis of a given area implies knowledge of the evolution of the relief, which can be obtained by studying the forms and successive deposits of preserved materials resulting from the different morphogenic processes to which the relief has been subjected. Casseti (1994) considers that the object of study of geomorphology is the surface of the earth's crust, which has a specific way of analysing relief. This type of analysis encompasses knowledge of the antagonistic forces provided by tectogenetic activities and morphoclimatic mechanisms responsible for the resulting forms present on the planet's surface.

1.2 Hydrogeomorphology

Goerl et al., (2012, p.105) states that Hydrogeomorphology deals with the aspects of water, rocks and morphological surface features responsible for the delineation of the landscape and includes the interrelationships between various hydrological and geomorphological processes and can be defined as an interdisciplinary science that is found in the connection and interaction of geomorphological processes with surface or subsurface water on different spatial and temporal scales. Okunishi (1994) explains that the flow of water over a moving surface on a sloping hillside or the bottom surface of a channel causes the movement of sediment and, consequently, topographical changes which, in turn, will control the characteristics of the water flow. The hydrodynamic focus refers to the energy and dynamics of water flow within the environment and the geomorphological focus deals with the landscape and its geological evolution and the primary source of water, which can be precipitation, surface water or groundwater (BRINSON, 1993). Extraordinary precipitation events trigger most landslides, but at the same time, the vast majority of slopes do not fail. What is difficult to say precisely is where, exactly, a slope can be triggered to slide and flow downhill (Boogard & Greco, 2015).

Beutler et al., 2001, p.169) consider that soil permeability depends, among other factors, on the quantity, continuity and size of the pores. Compaction and the discontinuity of these pores are responsible for significantly reducing the soil's permeability to water, but when there is fissuring, water can move relatively easily through the fractures. This can be seen in cases where a large amount of rain falls in a short space of time, causing floods capable of causing disasters.

1.3 Slope Instability Processes

The slope of a slope corresponds to the degree of inclination that the site has in relation to a horizontal axis, i.e. steeper slopes have a greater slope and the steeper a slope, the greater the risks of erosion processes becoming more pronounced, as well as gravitational mass movement (Vargas, 2015, p.79).

Zentner (2018) states that slopes undergo continuous variations represented by changes in their shape, and the steepest slopes demarcate an area in the relief that is unsuitable for human use or occupation, and the maximum limit for this is restricted to 30% inclination because these are places with a certain natural risk of gravitational mass movement.

The study of the geomorphology of slopes is of great importance because their improper occupation without proper planning can trigger natural disasters such as the occurrence of mass movements and large-scale erosion processes that cause material damage and loss of life. The instability of a slope can be accelerated by anthropogenic action on its natural process 64

altering the natural dynamics of the system. The steeper and more unvegetated the slopes, the greater the risk of landslides (Penteado, 1983, p.99).

The migration of large numbers of people to the coastal area in search of a better quality of life in past centuries characterised the appropriation and transformation of hillsides as property, implying spontaneous use and, consequently, a relationship of predation (Casseti, 1995). Guerra (2003) refers to the anthropogenic impact on natural slopes as the main factor influencing the processes, shapes and evolution of slopes.

Azevedo & Dalmonin (2004) point out that the inappropriate use of the slope environment through practices such as building roads and motorways on unstable soils, causing vibrations, excavating and opening trenches in rocky areas, as well as building industries and commercial areas on steep slopes without prior recognition of the soil resources that alter the natural dynamics of the landscape can result in mass movements and erosion processes.

A slope is considered stable if the force exceeds the denudation processes, and unstable if the materials are weaker than these processes. These erosion and mass movement processes occur on slopes, since the steeper the slope, the greater the chances of these processes occurring, as well as improper land use (Cristopherson, 2012).

Analysing the slope makes it possible to see the distribution of slopes on the surface of the land and is an important characteristic in analysing the balance of deposition and removal of material. The steeper the slope, the faster the potential energy of rainwater is transformed into kinetic energy, increasing the speed of the water and its transport capacity, which is responsible for the erosion processes that shape the landforms (Cristopherson, 2012).

In areas with erosion, the landscape is almost entirely formed by slopes, with the exception of the deepest valley regions. The slopes are in a dynamic equilibrium adjusting to the new surface conditions, i.e. the gravitational mass movements that represent the way they adjust to the conditions imposed by the change in slope gradient, water saturation, change in vegetation cover, overloading of the site and disintegration of material by the action of the weather (Guerra, 2011, p.150).

For Casseti (1994), among the processes that act on the earth's surface, denudation is of fundamental importance in the dynamics of the evolution of the relief, since its main action is the wear and reorganisation of the relief, making the energy of the relief one of the main modifiers of the environment, with weathering, erosion, transport, deposition and mass movements being considered the main processes of denudation.

1.4 Declivity Considerations

Casseti (1994) considers that the slope factor plays an important role in the occurrence of gravitational mass movements, since material exposed on sloping surfaces tends to be more influenced by the action of gravity. Beutner & Gerber (2005) refer to the variations suffered by slopes due to their gradient and point to the high concentration of rainwater in a given area, water discharged directly into the slopes, as well as the occupation and exposure of land susceptible to erosion. With regard to soil instability, it is difficult to determine exactly where the surface irregularities are, which is a problem for geomorphology.

The predominance of steep slopes on the earth's surface is the basis for countless cases of potential instability in terms of mass movement processes, given that occupations are recurrent in these areas. The instability of slopes is a natural process in which man's actions have altered the natural dynamics of the system (Penteado, 1983, p.99).

Guerra (2011) emphasises that relief is one of the elements that affects soil formation, as can be seen in the case of slopes that are too steep for the soil to fully develop, because gravity and erosion processes remove surface materials, including nutrients and minerals.

1.5 Mass Movements

Extreme weather events have become increasingly intense and frequent as a result of climate change in the world and in Brazil. In the last two decades, the states of Sao Paulo, Rio de Janeiro, Alagoas, Pernambuco, among others, have been intensely affected by extreme weather events that have caused floods, torrents and mass movements of various categories. These phenomena have had catastrophic consequences with a high number of economic losses and loss of human life (Highland & Brobowsky, 2008, p.3).

The type of landslide will determine the potential speed of movement, the likely volume of displacement, the distance of displacement, as well as the possible effects of the landslide and the appropriate mitigation measures to be considered. Highland & Brobowsky (2008, p.25) state that the material in a sliding mass can be rock, soil or both. The type of movement describes the internal mechanics of how the mass is displaced: falling, spanning, sliding, spreading or flowing.

2 METHODOLOGY

Google sites were used as a search engine to obtain Google Earth images, as well as information from academic articles, blogs, books and YouTube, from which images were captured of rockslides, debris and mud in some regions, where rainfall exceeded the historical average, causing environmental disasters with losses. The authors Varnes, (1978) and (Highland & Bobrowsky), 2008, who are considered by the academic world to be authoritative on the subject, were used to structure this work.

3 RESULTS AND DISCUSSION

The inability of scientists to prevent this type of accident because they don't know the structure of the soil and the permission to build houses on the bottom has been a major factor in the occurrence of this type of environmental disaster. Many people were killed in this landslide in the city of El Salvador. You can see the extensive layer of loose soil, which is highly prone to collapsing without any hindrance.

Figure 1 Barrier slide in El Salvador/2001
Fonte\; ElSalvadorslide.jpg
https://sv.wikipedia.org/wiki/Jordb%C3%A4vningama i El Salvador 2001#/media/Fil:ElSalvadorslide.jpg

In Figure 2, the arrow shows the cut made in the land to create a vehicular passage in the middle of the elevation, removing the support of the volume of earth at the top which, under the action of rain and gravity, ended up giving way and causing several houses to be buried with losses in southern California.

Figure 2 Landslide in La Conchita/Southern California
Source: Pubs.usgs.gov/2004
https://pubs.usgs.gov/fs/2004/3072/fs-2004-3072.html

Type of Movement	Type of material		
Falls	Rocky base	Engineering: Soil	
Tombo	Stone Fall	Predominantly	Predominantly

	Rockfall	Rock	Thin
Rotational	Slipping Rock	Drop in De tri to s	Land clearing
Translational		Overthrow Debris	Slipping debris
Lateral Spreading	Rock Propagation	Slipping debris	Slipping Earth
Flows	Rock flow (Deep Fluency)	Debris Flow	Earth flow
ComplexCombination	of two or more types of movement		

Table 1 Types of landslides. Adapted from the Classification of Slope Movements (Varnes, 1978, p.11).

Painting 1. Types of landslides. Adaptation of the classification of slope movements (Varnes, 1978, p.11).

The saturation of hillsides by water after heavy rains is the main cause of landslides. Other causes include snow melt, changes in groundwater levels and changes in water levels along coastlines, inland reservoirs and on the shores of lakes, reservoirs, canals and rivers (Highland & Bobrowsky, 2008, p.26).

Tumbling has recently been recognised as a distinct type of movement and consists of the forward rotation of one or more units about some pivotal point under the action of gravity and forces exerted by adjacent units or by fluids in crevices (Varnes, 1978, p.12). In translational sliding, the mass moves outwards or downwards along a gently undulating plane with little rotational or backwards movement, characteristic of the inclination of the fall. The moving mass generally slides on the surface of the ground (Varnes, 1978, p.13). The most common examples of rotational slides are small deformations, which slide along a rupture surface curved concavely upwards. They can be combined with other types of movement and make up a high proportion of landslide problems faced by engineering (Varnes, 1978, p. 13). In lateral spreading, the dominant mode of movement is lateral extension accommodated by shear or traction fractures (Varnes, 1978, p.14). Unconsolidated materials generally take the form of fairly common flows, fast or slow, wet or dry. In rock, the easiest movements to categorise include those that are extremely slow and distributed between many non-interconnected, closely spaced fractures, or movements within the rock mass that result in folding, bending or bulging (Varnes, 1978, p.17). The movements are generally extremely slow and can result in folding, bending, bulging or other manifestations of plastic behaviour (Varnes, 1978, p.17). The type of movement describes the internal mechanics of how the mass is displaced: falling, swaying, slipping, spreading or flowing, so landslides are described by the use of two terms that refer respectively to the type of movement and the quality of the material, i.e. rockfall or debris flow (Highland & Bobrowsky, 2008).

3.1 Rockfall

Rockfalls are the sudden downward movements of rock, earth or both that come off steep slopes or cliffs. The falling material usually hits the lower walls of the slope at an angle smaller than the angle of the fall, causing bouncing. The falling mass may break on impact or it may begin to roll down steeper slopes and may continue to the lower ground (Highland & Bobrowsky, 2008).

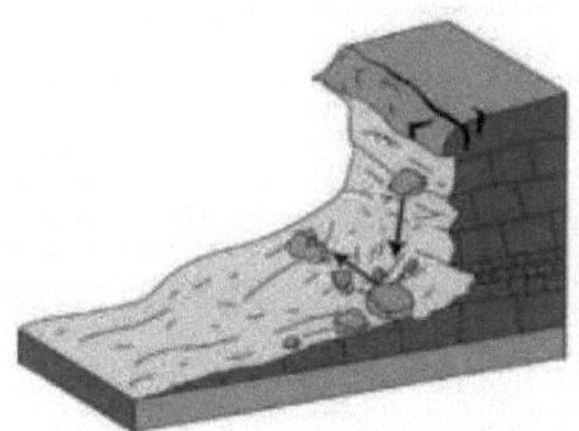

Figure 3 Rockfall diagram

Source: (Highland & Bobrowsky, 2008, p.27).

Figures 4, 5, 6 and 7 show images taken from a youtube video that faithfully express the descriptions provided by (Highland & Bobrowsky, 2008). The images represent one of the greatest dangers for those who live and travel in areas surrounded by large elevations formed by rock and earth. Not even the trees below can contain the displacement, and the force of the rocks increased by gravity crushes everything in front of it. Located on a very rugged terrain, the northern state of Sikkim has heavy and well-distributed rainfall from May to early October. July is the wettest month in most places, but very high rainfall amounts have occurred in August and September, as shown in Table 1. The highest annual rainfall for each season can exceed 5000 mm. During the mongoons, heavy rainfall increases the risk of landslides (Sikk geo, 2019). The year 2019 has been the scene of countless calamities around the globe, and the excessive volume of rainfall has so far fallen without explanation in a delimited area, causing disasters both inside and outside the urban perimeter. The amount of rainfall is totally out of the ordinary when it exceeds the historical average, raining in one day what was expected for a month, as was the case in the municipality of Camaragibe in Pernambuco, where the amount of rainfall reached 177 mm in 24 hours (G1PE, 2019). On 11 January 2011, it rained 400 mm in just one night in the city of Nova Friburgo. Saudi Arabia had its desert transformed into a huge lake after it rained the amount expected for six years in just a few days.

3.2 Tax liens

A toppling is identified by the forward rotation of a mass of soil or rock out of the slope, around a point, or axis, below the centre of gravity of the displaced mass. Overturning is sometimes caused by gravity exerted on the weight of the material at the top of the displaced mass. Slumps can contain rocks, debris (coarse-grained material) or earth (fine-grained material). They can also be complex and compound. Sometimes the overturning is due to water or ice in the cracks of the mass (Highland & Bobrowsky, 2008 p.29).

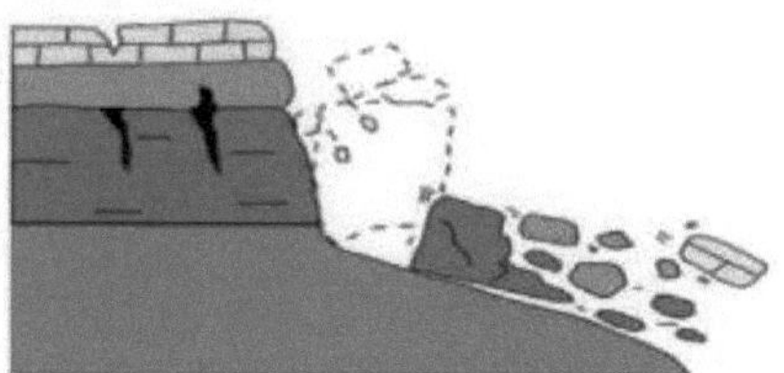

Figure 8 Listing Scheme by Highland & Bobrowsky, 2008, p.30

Figura 1 1 Debris Flow Imacio
https://www.youtube.com/watch?v=FatbfDdQhgE. August/2013/Taiwan

Figura 2 2 Dumping of Debris in the Boulevard
https://www.youtube.com/watch?v=FatbfDdQhgE. August/2013/Taiwan

Figura 3 3 Rockfall in a Busy Area
https://www.youtube.com/watch?v=FatbfDdQhgE. August/2013/Taiwan

Figure 10 Truck stuck in debris Figure 10.1Splitting of rocks and debris

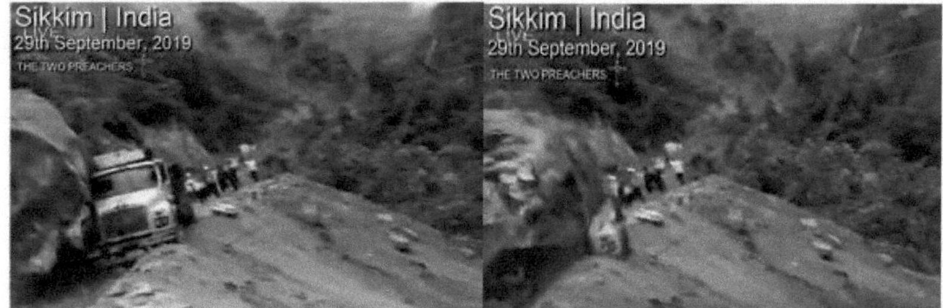

Figure 10.3 Rock detachment Figure 10.4 Rock crushing the lorry

3.3 Rotational slip

It is a type of sliding in which the rupture surface is curved upwards in the shape of a spoon and the movement of the barrier drop is more or less rotational around an axis parallel to the contour of the slope. The displaced mass can, under certain circumstances, move relatively coherently along the rupture surface and with little internal deformation. The top of the displaced material can move almost vertically downwards and the top of this material can tilt back towards the slope. If the slip is rotational and has several parallel and curved planes of movement, it is called subsidence (Highland & Bobrowsky, 2008, p. 31).

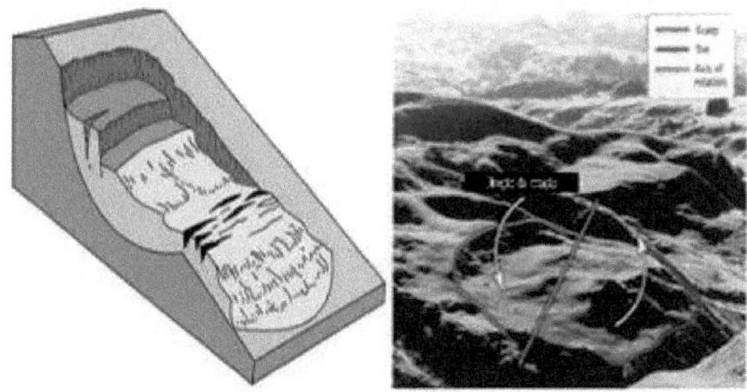

Figure 11a Rotational Slip Diagram Figure 11b Slip

Source: Highland & Bobrowsky, 2008, p. 32

3.4 Translational slippage

The mass of a translational slip moves outwards, or downwards and outwards along the relatively flat surface, with little rotational movement or backwards inclination. This type of slip can progress for considerable distances if the rupture surface is sufficiently inclined, unlike rotational slips, which tend to restore slip equilibrium. The material in the slide can vary from loose, unthickened soil to large slabs of rock, or both. Translational slides commonly occur along geological discontinuities such as faults, joints, surfaces, stratifications, or the point of contact between rock and soil. In places in the northern hemisphere, this type of sliding can also move along permanently frozen subsoil layers (Highland & Bobrowsky, 2008, p. 34).

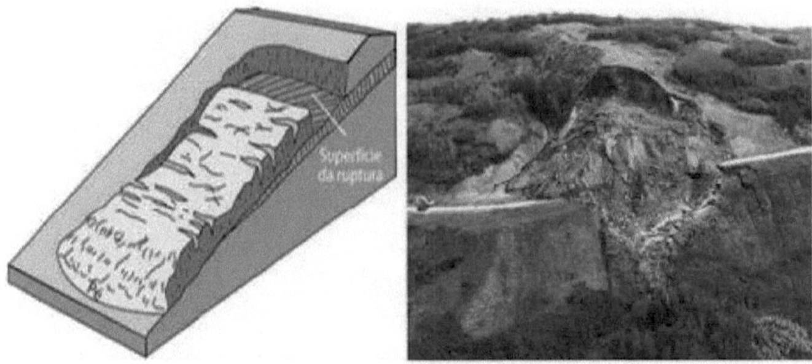

Figure 12a Slip Scheme Figure 12b Translational Slip
(Highland & Bobrowsky, 2008, p. 35).

3.5 Lateral Spreading

It occurs on slopes with a low inclination or on flat ground, especially where there is an upper layer of rock or soil which, as it increases in length, moves over a less steep and weaker layer. Such ruptures are often accompanied by a generalised subsidence towards the lower layer. Such ruptures are commonly accompanied by a generalised subsidence towards this lower layer. In rock spreading, the solid part extends and ruptures by slowly moving from its stable position to the top of the weaker layer, without necessarily forming an identifiable rupture surface. In land spreading, the stable upper layer extends over the weaker lower layer, which flows after liquefaction or plastic deformation (Highland & Bobrowsky, 2008, p. 36).

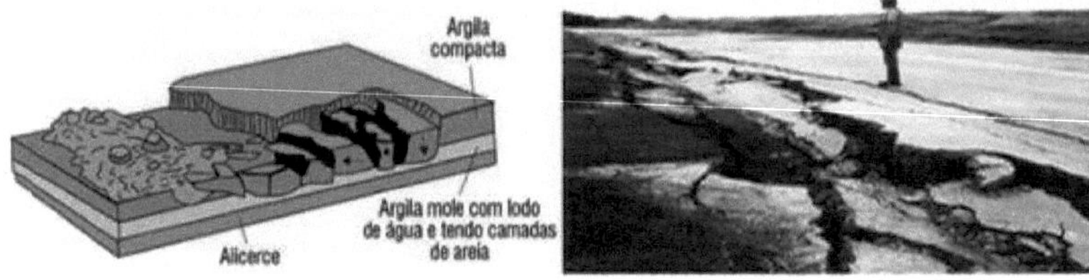

Figure 13a Schematic of Lateral Spreading Figure 13b Lateral Spreading-USA
Source: (Highland & Bobrowsky, 2008, p. 39).

3.6 Flow

Flow is a spatially continuous movement in which the shear surfaces are short-lived, have little space between them and are not preserved. The amount of water present will influence the mobility and evolution of the movement.

3.7 Debris Flow

It is a rapid form of mass movement, in which loose soil and rocks, and sometimes organic material, mix with water to form mud that flows down the slope and can cause fatalities, as it is extremely fast and can occur without any warning (Highland & Bobrowsky, 2008, p. 40).

Figure 14a Flow diagram Figure 14b Flow in Venezuela (Highland & Bobrowsky, 2008, p.

42).

3.8 Debris avalanche

Debris avalanches are essentially large and extremely fast-moving, often forming flows on open slopes and are formed when an unstable slope breaks and the resulting debris is rapidly transported away. In some cases, snow and ice will contribute to the movement if there is enough water and the movement can become a debris flow (Highland & Bobrowsky, 2008, p. 46).

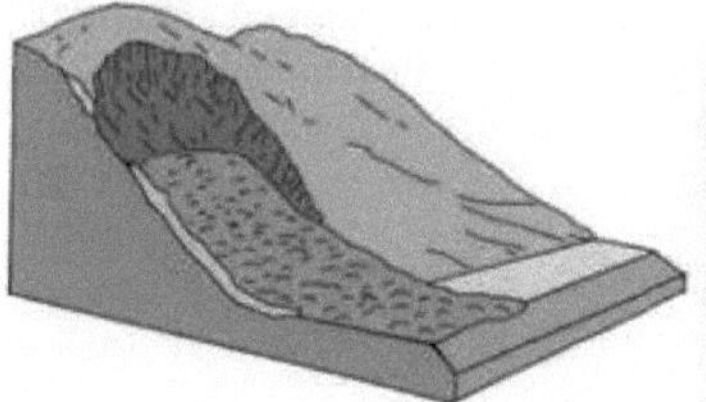

Figure 15ª Debris Avalanche Scheme Figure 15b Debris Avalanche
(Highland & Bobrowsky, 2008, p. 47)

The image in Figure 15b shows the debris avalanche that buried the village of Guinsaugon, south of Leyte, Philippines in February 2006.

3.9 Crawl

Also called slow earth flow, creep is the name given to slow earth flow and consists of a slow, imperceptible and continuous downward movement of the soil that forms the slope. This type of displacement is caused by internal shear stress that is sufficient to cause deformation but insufficient to cause rupture. Creep is everywhere on the planet and is probably the most common type of landslide, often preceding other, faster and more damaging types of landslides (Highland & Bobrowsky, 2008, p. 51).

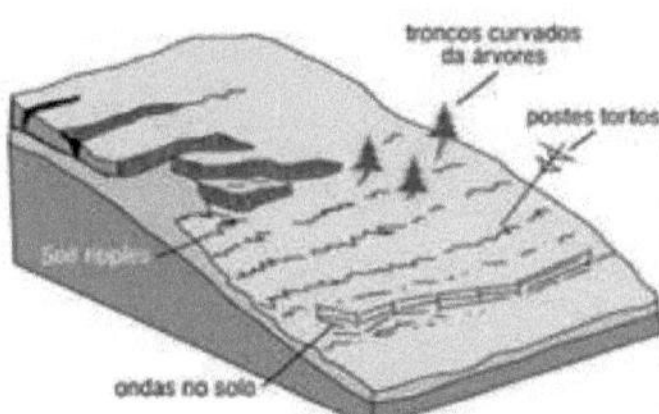

Figure 16a Crawling diagram Figure 16b Crawling in the UK
(Highland & Bobrowsky, 2008, p. 52-3).

3.10 Increase in Global Precipitation

Water saturation on slopes is the main cause of landslides. Saturation can occur in the form of heavy rainfall, thawing, changes in water movement in the soil, changes in water movement on the surface along coastlines, earth dams and on the shores of lakes, reservoirs, canals and rivers. Landslides and floods are closely associated, as both are related to precipitation, runoff and soil saturation by water (Highland & Bobrowsky, 2008, p. 59). In the last two decades, the volume of precipitation has increased significantly and is the main cause of the landslides that have claimed so many victims, mainly due to difficulties in monitoring risk areas. When these houses were built, the amount of rainfall was not enough to cause disasters such as the burial of houses. In a study carried out in Australia, Alexander et al. (2006) state that rainfall rates have tended towards wetter conditions throughout the 20th century. Westra et al. (2013) found results that indicate statistically significant upward trends that can be detected on a global scale, with almost two-thirds of the seasons showing an increase. Meehl et al. (2007, p.750) confirmed that precipitation extremes increase more than average in more tropical mid- and high-latitude areas. Studies have shown that precipitation has tended to increase in the mid-latitudes more towards the southern hemisphere than towards the north (Folland et al., 2018). Blunden & Arndt (2017), studying the 1901-2000 series as a baseline to represent changes, conclude that the total annual amount of precipitation over land around the world has changed. Lechmann et al., (2015) studied record rainfall events between 1981 and 2010 and concluded that the number of these events has increased significantly in the global average. Trembert et al., (2007) studied the historical series from 1900 to 2005 and concluded that heavy rainfall events increased more in the northern hemisphere. Donat et al. (2013), using observational data, conclude that precipitation extremes and annual precipitation totals have increased over the last sixty years.

Figures 17 to 18 show the houses in the town of Nova Friburgo built on the base of the hills with vegetation and clay soil that received a total of 400mm of rain in one night on 11 January 2011, which caused the disaster.

Figure 17a Praça do Suspiro Figure 17b City Centre
Source: https://www.youtube.com/watch?v=lgXzielFuFg

Figure 18 a Sao Lucas Hospital Figure 18b Corrego Dantas
Source: https://www.youtube.com/watch ?v=lgXzielFuFg

Figure 26 a and b Mud avalanche in Chamonson-Suiga/2018
Source: https://www.youtube.com/watch?v=2kTzJdj8BDI

4 FINAL CONSIDERATIONS

Landslides have become one of the most common disasters during the rainy season, particularly in hillside areas in regions with very rugged terrain, as is the case of the cities located in the mountainous region of Rio de Janeiro. A few decades ago, here in Brazil, municipal governments were concerned about the fragility of the hills in the Southeast and Northeast, but today, environmental disasters caused by landslides have occurred in many countries, including those considered to be in the first world. Environmental scientists have a hard time finding out where the event will occur, mainly because they can't measure the bearing capacity of the soil. One factor that has been occurring before the landslides is the increase in rainfall in a short space of time, when it rains in a few days uninterruptedly, what should rain in six years, as is the case in Saudi Arabia in 2019, whose desert turned into a river of water and mud causing many victims. Here in Brazil, in the mountainous region of Rio de Janeiro in 2011, it rained in one day what was expected for the whole month and the result was catastrophic. The government has employed slope stabilisation techniques, but families continue to insist on living in the danger zones in the belief that high rainfall events will not happen again. This phenomenon of dumping a large amount of rain in a single place in a short space of time is very difficult to prevent, so in view of the forecast, people living in these areas should look for safer places.

5 REFERENCES

Alexander, Lisa. Zhang, Xuebin, Peterson, T C. Caesar, J. Global observed changes in daily climate extremes of temperature and precipitation. 1, 198. March. 2006. DOI: 10.1029/2005JD006290. Available at: https://www.researchgate.net/publication/43155544 Global Observed Changes in Daily Climate Extremes of Temperature and Precipitation. Accessed on 16/10/19 at 14:49hs.

Brazil's biggest natural disaster - 400mm of rain in one night. 11 January 2011. Nova Friburgo tragedy seen from space. Landslide in Praga do Suspiro with burial. Available at: https://www.youtube.com/watch?v=lgXzielFuFg. Accessed on 03/10/2019 at 22:28.

Mud avalanche in Suiga scares residents. 08/08/2018.Chamonson-Suiga. Available at: https://www.youtube.com/watch?v=2kTzJdj8BDI. Accessed on 17/10/2019 at 17:46hs.

AZEVEDO, A. C.; DALMOLIN, R. S. D. **Soils and the environment: an introduction.** Santa Maria: Palloti, 2004. In: Pedron, Fabricio de Araujo. Dalmolin, Ricardo Simao Diniz. Azevedo, Antonio Carlos de. Botelho, Marcio Ramos. Menezes, Fabio Pacheco. Soil Survey and Classification in Urban Areas: Importance, Limitations and Applications. R. bras. Agrociencia, Pelotas, v.13, n.2, pp.147-151, abr-jun. 2007.

BBC Brasil - Car escapes giant rock that fell onto road - 31 August 2013 https://www.youtube.com/watch?v=FatbfDdQhgE 17/10/2019 at 15:41hs.

Beutler, A N. Silva, M L N. Curi, N. Ferreira, M M. Cruz, J C. Pereira Filho, I A. Penetration resistance and permeability of typical dystrophic red latosol under management systems in the cerrado region. R. Bras. Ci. Solo, 2001,25:167-177. DOI: http://dx.doi.org/10.1590/S0100-06832001000100018.

Blunden, J. Arndt, D S. State of the Climate. Bulletin of the American Meteorological Society. Vol.98, n.8, August. 2017. 298p. DOI: 10.1175/2017.

Bogaard, T. Greco, Roberto. Landslide hydrology: from hydrology top ore pressure. In: Wiley Interdisciplinary Reviews: Water. 3(3):n/a-n/a. November 2015. DOI: 10.1002/wat2.1126

Brinson, M M. A Hydrogeomorphic Classification for Wetlands. Wetlands Research Programme Technical Report. WRP-DE-4, U.S. Army Corps of Engineering, 1993. 101p.

Casseti, Walter. Elements of Geomorphology. In: Vargas, Luciani Vieira de. Susceptibility to mass movements: a geomorphological study of the Vacaca^ Mirim river sub-basin upstream of the DNOS dam in Santa Maria/RS. Master's dissertation. 132p. UFSM-Centre of Natural and Exact Sciences. Santa Maria, RS, Brazil. 2015.

Christopherson, Robert. Geosystems: an introduction to Physical Geography. In: Vargas, Luciani Vieira de. Susceptibility to mass movements: a geomorphological study of the Vacaca^ Mirim river sub-basin upstream of the DNOS dam in Santa Maria/RS. Master's dissertation. 132p. UFSM-Centre of Natural and Exact Sciences. Santa Maria, RS, Brazil. 2015.

District Rainfall (mm) North Sikkim Region 2009/2018. Source: http://sikenvis.nic.in/WriteReadData/UserFiles/file/North%20Sikkim%20%202009-2018.pdfAccessed on: 13/10/2019 at 12:10hs.

Rock and Mudslide: Bondasca Valley bordering Italy. Available at: https://www.youtube.com/watch?v=3E2PdAswK44. Accessed on 03/10/2019 at 23:09.

Donat, M G., et al. Update analyses of temperature and precipitation extreme indices since the beginning of the twentieth century: The HadEX2 dataset, J. geophys. Res. Atmos. 118, 2098-2118. 2013. DOI: 10.1002/jgrd.50150.

Barrier slide in El Salvador/2001. Available at: https://sv.wikipedia.org/wiki/Jordb%C3%A4vningarna i El Salvador 2001#/media/Fil:ElSalvadorslide.jpg. Accessed on: 04-10-2019 at 01:03hs.

Florenzano, Teresa Galloti. Geomorphology: current concepts and technologies. In: Vargas, Luciani Vieira de. Susceptibility to mass movements: a geomorphological study of the

Vacaca^ Mirim river sub-basin upstream of the DNOS dam in Santa Maria/RS. Master's dissertation. 132p. UFSM-Centre of Natural and Exact Sciences. Santa Maria, RS, Brazil. 2015.

Folland, C K Karl, T R Vinnikov, Y A. Observed Climate Variations and Change. Chap.7. 44p. Available at: https://www.ipcc.ch/site/assets/uploads/2018/03/ipcc_far_wg_I_chapter_07-1.pdf. Accessed on 16/10/2019 at 15:31hs.

G1 PE - Barrier collapse kills man and leaves people missing in Camaragibe 13/06/2019 15:49 Updated 3 months ago. Available at: https://g1.globo.com/pe/pernambuco/noticia/2019/06/13/queda-de-barreira-mata-homem-_e-deixa-pessoas-desaparecidas-em-camaragibe.ghtml Accessed on 05/10/2019 at 15:49hs. 21:31 on 05-10-2019

Goerl, R F. Kobyama, Massato. Santos, Irani dos. Hydrogeomorphology. Brazilian Journal of Geomorphology. V.13, n.2, 2012, Apr-Jun,pp.103-111.

Goudie, Andrews S. Encyclopedia of geomorphology. Volume 1. 1202p.London: Routledge. 2004. Available at: https://courses.ess.washington.edu/ess306/links/Goudie_Encyclopedia_of_Geomorphology.pdf.

Accessed on: 18/10/2019 at 12:18 pm.

Guerra, Antonio Jose Teixeira. Urban Geomorphology. In: Vargas, Luciani Vieira de. Susceptibility to mass movements: a geomorphological study of the Vacaca^ Mirim river sub-basin upstream of the DNOS dam in Santa Maria/RS. Master's dissertation. 132p. UFSM-Centre of Natural and Exact Sciences. Santa Maria, RS, Brazil. 2015.

Highland, Lynn M. Bobrowsky, Peter. The Landslide Handbook - A Guide to Understanding Landslides Lynn M. Highland United States Geological Survey Peter Bobrowsky Geological Survey of Canada. USGS Science for a changing world. 176p. Available at: https://www.gfdrr.org/sites/default/files/publication/Deslizamentos_M5DS_0.pdf

Accessed on 18/10/2019 at 11:13 am.

La Conchita, coastal area of southern California. This landslide and earthflow occurred in the spring of 1995. People were evacuated and the houses nearest the slide were completely destroyed. This is a typical type of landslide. Photo by R.L. Schuster, U.S. Geological Survey. Available at: https://pubs.usgs.gov/fs/2004/3072/fs-2004-3072.html Accessed on 04/10/19 at 01:07hs.

Lechmann, Jascha. Coumou, Dim. Frieler, Katja. Increasing record-breaking precipitation events under global warming. Climate Change. 132(4). August. 2015. DOI: 10.1007/s10584-015-146-3.

Okunishi, K. Concept and methodology of hydrogeomorphology. Transactions, Japanese Geomorphological Union, 15ª , pp. 5-18. 1994.

Penteado, Margarida Maria. Fundamentals of Geomorphology. In: Vargas, Luciani Vieira de. Susceptibility to mass movements: a geomorphological study of the Vacaca^ Mirim river sub-basin upstream of the DNOS dam in Santa Maria/RS. Master's dissertation. 132p. UFSM-Centre of Natural and Exact Sciences. Santa Maria, RS, Brazil. 2015.

Popp, Jose Henrique. General Geology. In: Vargas, Luciani Vieira de. Susceptibility to mass movements: a geomorphological study of the Vacaca^ Mirim river sub-basin upstream of the DNOS dam in Santa Maria/RS. Master's dissertation. 132p. UFSM-Centre of Natural and Exact Sciences. Santa Maria, RS, Brazil. 2015.

Sikk Geo. Geographical Information of Sikkim. Available at: http://sikkim.nic.in/sws/sikk_geo.html. Accessed on 13-10-2019 at 12:02 pm.

Stetler, Larry. Geomorphology. Reference Module in Earth Systems and Environmental Science. 1-10. 2014. DOI: 10.1016/B978-0-12-409548-9.09078-3. Available at:

https://www.researchgate.net/publication/260363052 Geomorphology.Accessed on 17/10/2019 at 22:10hs.

Tavenas, F. Chagnon, JY. Rochelle, P. 1971. The Saint-Jean-Vianney. Landslide: Observations and Eyewitness Accounts. Canadian Geotechnical Journal. 2011. 8(3):463-478. DOI: 10.1139/t71-048. Retriev 22/07/2010.

Trenberth, K E. Jones, P. D. Ambenje, R. Bojariu, D. et al. Observations: Surface and Atmospheric Climate Change. Chapter 3. In: Climate Change 2007: The Physical Science Basis. 102p. Available at: https://www.ipcc.ch/site/assets/uploads/2018/02/ar4-wg1-chapter3-1.pdf. Accessed on: 16/10/2019 at 21:43 hs.

Torres, Fillipe Tamiozzo Pereira. Neto, Roberto Marques. Menezes, Sebastiao de Oliveira. Introduction to Geomorphology. In: Vargas, Luciani Vieira de. Susceptibility to mass movements: a geomorphological study of the Vacaca^ Mirim river sub-basin upstream of the DNOS dam in Santa Maria/RS. Master's dissertation. 132p. UFSM-Centre of Natural and Exact Sciences. Santa Maria, RS, Brazil. 2015.

Vargas, Luciani Vieira de. Susceptibility to mass movements: a geomorphological study of the Vacaca^ Mirim river sub-basin upstream of the DNOS dam in Santa Maria/RS. Master's dissertation. 132p. UFSM-Centre of Natural and Exact Sciences. Santa Maria, RS, Brazil. 2015.

Varnes, David J. Slope Movement Types and Process. Chapter 2, 23p. Available at: http://onlinepubs.trb.org/Onlinepubs/sr/sr176/176-002.pdf . In: https://trid.trb.org/view/86168. Accessed on 17/10/2019 at 10:58 am.

Westra, Set. Alexander, Lisa. Global Increasing Trends in Annual maximum Daily Precipitation. Journal of Climate. 2(11):7834. April. 2013. DOI: 10.1175/JCLI-D-12- 00502.1.

Wicander, Reed. Monroe, James. Fundamentals of Geology. In: Vargas, Luciani Vieira de. Susceptibility to mass movements: a geomorphological study of the Vacaca^ Mirim river sub-basin upstream of the DNOS dam in Santa Maria/RS. Master's dissertation. 132p. UFSM-Centre of Natural and Exact Sciences. Santa Maria, RS, Brazil. 2015.

Zentner, Nick. Bridge of the Gods landslide. Central Washington University. Retrieved 23 December 2018. Available at http://www.nickzentner.com/https/youtube/mxeqetg0xww. Accessed on 17/10/219 at 20:15.

CHAPTER 5

EXTREME HYDROLOGICAL EVENTS: REFLECTIONS ON THE WATER HEAD PHENOMENON

The "headwater" phenomenon is an extreme hydrological event characterised by a sudden rise in the water level in rivers, streams or waterfalls, usually occurring in mountainous areas. This phenomenon is the result of intense rainfall at the headwaters of rivers, which may not be visible in the lower areas where the flood occurs, taking many by surprise. Here are some thoughts on the phenomenon:

Characteristics of the Cabega d'Agua Phenomenon:

1. **Origin and causes**:

This is caused by torrential rains that occur in the upper parts of a river basin. These heavy rains quickly saturate the soil, resulting in accelerated surface runoff, which flows down the slopes and reaches the watercourses in the lower areas with great force.

2. **Rapid Rise in Water Level**:

o One of the most dangerous aspects of "cabega d'agua" is the speed with which the river or stream rises. In a matter of minutes, the water can rise several metres, creating a strong and unexpected current with the potential to sweep away people, vehicles and destroy infrastructure.

3. **Unpredictability**:

o The phenomenon is notoriously difficult to predict, especially as the rainfall that causes it can occur far from the affected area. Thus, even on an apparently sunny day at a waterfall or river, a "water head" can occur due to rainfall in distant areas.

Impacts and Risks:

1. **Risks to Human Life**:

Due to its sudden nature, "cabega d'agua" represents a major risk for bathers, fishermen and people who live or work near rivers and waterfalls. Unfortunately, there are many cases of deaths and disappearances associated with this phenomenon.

2. **Environmental Impact**:

o In addition to human risks, "water logging" can cause significant erosion of river banks, alter aquatic habitats and impact local fauna. It can also transport large quantities of sediment, polluting water bodies and affecting water quality.

3. **Damage to infrastructure**:

o The force of the water can damage bridges, roads and other infrastructure located near watercourses. In tourist areas, the economic damage can be substantial, due to the closure of parks and recreational areas after a "head of water".

Prevention and Mitigation Measures:

1. **Monitoring and Warning**:

o Implementation of hydrological monitoring systems that can quickly detect changes in river flow and issue alerts to downstream areas. This could include meteorological stations, rain radars, and alert systems via SMS or mobile applications.

2. **Education and Awareness**:

o Informing the public, especially in tourist areas, about the risks of the phenomenon and how to recognise signs of a "water bollard". This includes educational campaigns and appropriate signposting in prone areas.

3. **Risk Management**:

o Develop specific emergency plans for vulnerable areas, including escape routes, safe meeting points and evacuation procedures.

4. **Engineering and Infrastructure**:

o Design and build infrastructure, such as bridges and overpasses, that can withstand the impact of a "water head". In some areas, it may be necessary to prohibit construction or activities in risk zones.

The "water head" is an example of how extreme hydrological events, intensified by climate change, can pose significant and unpredictable risks. The combination of effective monitoring, education and resilient infrastructure is crucial to mitigating the impacts of this natural phenomenon and protecting lives and property. In recent decades, extreme rainfall events have been occurring on a global scale in countries on different continents and at the same time, drawing the attention of scientists and scholars with regard to the variations observed in the climate. This research aims to make an association between extreme rainfall events and the emergence of a phenomenon known in Brazil as Cabega D'Agua, which is characterised by heavy rainfall in a given region over a short period of time. It was observed that the phenomenon occurred between the months of July, August and September 2021 in countries on all continents. We focused our analysis on the catastrophic events that occurred in China, Belgium, Germany and Saudi Arabia in an attempt to synthesise the study, because dozens of countries were affected by the rains.

1 Introduction

Extreme weather events have been one of the main consequences of climate change in many places around the world. Extreme precipitation events are generally characterised by atypically high or low values recorded within a period of time and observation (SANTOS, 2014).

In the last decade, record rainfall has occurred in many places around the world, causing serious impacts on human society and the environment, including loss of life and agricultural losses due to flooding. Scientists continue to claim that human-induced greenhouse gases contribute to changes in temperature and precipitation events on a global scale. In the last three decades, the number of record precipitation events has increased significantly on a global scale (LEHMANN et al., 2015).

In the study of environmental vulnerability, one of the main causes is anthropogenic action in the formation of urban centres at the base of or near mountainous areas, which favours the receipt of abnormal rainfall discharges causing material losses and loss of life. It is important to verify the trend of abnormal rainfall events and seek to quantify them in order to develop tools and strategies to relocate homes to areas further away from rivers and mountains (NUNES et al., p.3, 2018).

Taking into account the impossibility of obtaining the duration of precipitation and, consequently, estimating the average intensity of events, Nunes et al., (p.6, 2018) obtained the median of daily precipitation, a measure of central tendency that is less susceptible to the presence of atypical values, making it possible to define a limit of 40mm, whose adopted reference value was checked for relevance to studies applied to the city of Belo Horizonte and other locations were analysed.

Chaves et al (1985) considers that rainfall is one of the main environmental factors that triggers erosion processes. Under natural conditions, soil loss increases with the increase in annual rainfall, which is always correlated with other factors such as slope and vegetation cover. Soil in urban areas near the base of mountains should be analysed for percolation, runoff and infiltration to prevent future disasters.

Thus, water and its various intervention processes modify and shape the earth's surface, together with other natural phenomena that occur throughout the evolution of the relief, which is made up of different geomorphological features that make up the different landscapes of the earth's surface.

Freitas (2021) states that in different parts of the world, precipitation is the most important factor in the formation of landforms, since rainwater results in torrents and, in some cases, landslides. The latter is due to the high amount of water contained in the soil, which leads it to enter a saturation stage. When it reaches this point, the soil gains too much weight, cannot support itself and breaks, causing a huge amount of soil to slide down the slopes. Often the soil doesn't absorb water because of the rapid movement of the torrents, so it doesn't cause the landslide, but cracks or gullies form on the surface. The gullies grow to larger proportions and become vogorocas (FREITAS, 2021).

1.1 The Hydrological Phenomenon of Cabega D'agua

In recent decades, a hydrological phenomenon has been occurring in several countries on several continents. In Brazil, this phenomenon is known as cabega d'agua, which is characterised by a sharp discharge of rain in a particular place, usually where the relief is composed of rocky mountains or not, covered with a lot of vegetation. Collischonn and Kobiyama (2019) describe this phenomenon as a type of flood in which the increase in flow at a given location is not just rapid, but practically instantaneous. In this type of flood, it is possible to clearly observe the arrival of the wave as a visible discontinuity in the flow and movement of the water, which can surprise people on the banks of streams, houses on the slopes of hills, inside cars on the streets, through which they are dragged for kilometres and can carry rocky sediments and woody debris that further enhance their lethality by generating debris flows or hyper-concentrated runoff.

The voluminous discharge of water down slopes can transport sediment, rock debris and large, medium and small vegetation. Azevedo (2021) refers to **flooding** as a meteorological phenomenon caused by the rapid and sudden increase in the water level in rivers or dry areas due to a large amount of rain falling in a given area. Analysing the reports of this type of flood, it is possible to conclude that water heads occur above all in small basins with steep slopes and soils with low infiltration capacity. In addition, water heads occur when the initial depth of the river, before the flood arrives, is low. The spatial distribution of the rain is concentrated in the headwaters or in the highest parts of the terrain and is capable of creating a wave that descends from the highest parts, sweeping away gravel, debris, trees, stones and mud.

The maximum flows and maximum water movements reported in headwater floods are not necessarily greater than other floods recorded in the same locations. Even so, headwater floods are characterised by such a rapid increase in water movement and flow that they can be life-threatening for people in the riverbed, such as bathers, animals, climbers and tourists, people inside houses, cars in the middle of the street or anywhere that is in the path of the water. For this reason, the greatest impacts associated with headwater floods are the loss of human life.

1.2 Hydrological Phenomenon in the Year 2021

1.2.1 Floods in Belgium

As a basis for comparison, in the year 2017 obtained from theglobaleconomy.com (2021) it rained the equivalent of 847mm. In 2021, the rainfall was heaviest in eastern Belgium, with 271.5 metres (10.69 in) of rain in 48 hours in the town of Jalhay, in the province of Liege - an absolute record for Belgium, almost three times the average rainfall in a month in this location for July. In the town of Spa, also in Liege, 217 mm (8.5 in) of rain fell over 48 hours. In Reifferscheid, 207 mm (8.1 in) fell within a nine-hour period, while Cologne saw 154 mm (6.1 in) in 24 hours.

Large parts of the province of Luxembourg saw between 150 mm and 200 mm in 48 hours. In Flanders, the maximum rainfall was 77 metres (3.0 in) over 48 hours at Ransberg. In Dilsen-Stokkem, Limburg, the river Maas (Meuse) reached a flow of 3300 m3 / s on 16 July, equalling the maximum flow that the local dykes were designed to withstand (Wikipedia.org, 2021).

Figures 1 and 2 Flood in Esneux - Belgium/2021
Source: The Guardian.com/Environment/jul/2021

In Figure 2, the blue arrow indicates higher areas where the volume of heavy rainfall is projected in large quantities over a short period of time, causing it to wash downhill, dragging away vegetation made up of trees and shrubs, as well as dry trunks and branches associated with mud. These elements are responsible for clogging up canals and knocking down houses along the way. This sudden descent of a large volume of water is known as a head of water, named after the states of the Southeast because they have mountainous areas covered in rocks, soil and abundant vegetation.

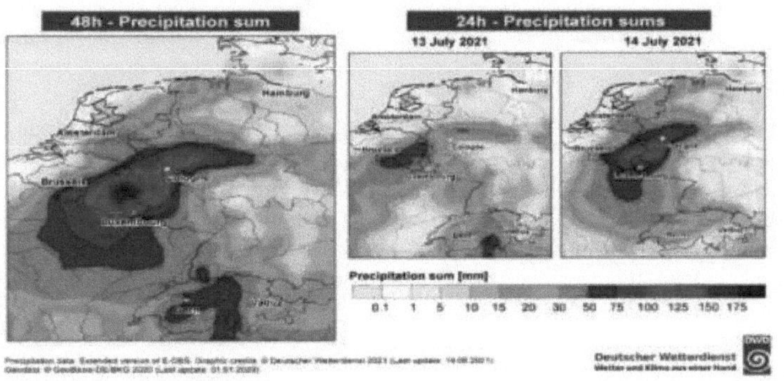

Figure 3 Precipitation scale in Belgium for the 24 and 48 hour values.
Source: DWD - Deutscher Wetterdienst

Figure 3 shows the concentration of rainfall in a particular region, which, if it is mountainous, will result in the formation of a waterfall. On 24 July, exceptionally strong thunderstorms resulted in almost 25,000 lightning strikes recorded between noon and 220pm, according to SRF Meteo. In Appenzell Innerrhoden (Suiga), 33.2 mm of rain fell in just 10 minutes, which is exceptionally heavy according to SRF Meteo. A particularly severe flash flood was reported in the London area, where 47.8 mm (1.88 in) of rain was recorded on 12 July at Kew, marking the third wettest day on record for that weather station and the wettest since 6 July 1983. Both Putney in London and Chipstead in Surrey recorded more than 31 mm (1.2 in) of rain in a one-hour period, while other areas of London recorded more than 76.2 mm (3.00 in) of rain in 90 minutes. The London Fire Brigade received more than 1,000 calls relating to flooding incidents, as homes were evacuated and cars submerged due to the rapid increase in flooding (Wikipedia.org, 2021).

Flooding in France According to Meteo-France, between Monday 12 July, 8:00 and Friday 16 July, 12:00, 199 mm (7.8 in) of rain fell in Chatel-de-Joux (Jura), 160 mm in Plainfaing (Vosges), 159 mm (6.3 in) in le Fied (Jura) and 158 mm (6.2 in) in Villiers-la-Chevre (Meurthe-et-Moselle) (Wikipedia.org/2021).

Although heavy rains remained occasional from the 7th to the 9th, the heaviest rains occurred from the 10th to the 12th of August, and several flood warnings were issued by the Dirección General de Meteorología. Some stations recorded rainfall of over 400 millimetres (16 inches) over these 48 hours, and numerical models suggested the possibility of a local maximum of up to 1,281 mm (50.4 inches), with hourly rainfall rates of well over 125 millimetres (5 inches) per hour. This probably caused severe flooding in Kastamonu, starting about a day after the second rainy season, when a nearby river burst its banks. Other regions also accumulated considerable rainfall, as Ayancik, Kure, Pinarbagi, Azdavay and Inebolu received 240 mm (9.4 in), 198 mm (7.8 in), 167 mm (6.6 in), 145 mm (5.7 in) and 123 mm (4.8 in) of rain respectively, mainly in the space of a few hours (Wikipedia.org, 2021). Since 17 July 2021, the Chinese province of Henan has been affected by severe flooding, caused by a period of intense and prolonged rainfall. A record maximum rainfall of 201.9 metres (7.95 in) was recorded in one hour in Zhengzhou, the provincial capital. Nineteen meteorological stations in the province renewed their daily rainfall records. On 2 August 2021, the provincial authorities reported that 302 people had died, with a further 50 missing, 815,000 people evacuated, 1.1 million relocated and 9.3 million people affected. Flooding has become more likely due to increases in extreme weather conditions caused by climate change in China (Wikipedia.org, 2021).

Zhengzhou

On 16 July 2021, Zhengzhou in China began to experience heavy rainfall. On 20 July alone, the average rainfall that day reached 253 mm (9.96 in). From 4pm to 5pm on 20 July, rainfall in one hour reached 201.9 mm (7.95 in); and from 8pm on 17 July to 8pm on 20 July, rainfall reached 617.1 mm (24.30 in) over three days.) over three days, (Aljazeera, 2021) close to the average annual rainfall (Washington Post, 2021). Videos emerged showing metro passengers waist-deep in water inside their carriages and cars floating in the streets (NBC, News, 2021).

Figure 4 Wion Climate Tracker, 2021 - Henan: The heaviest rainfall in 1000 years.
Source: https://www.youtube.com/watch?v=auCzJFxQbql

Henan was one of the regions of China hit hard by the rains, which caused an unprecedented catastrophe in urban centres.

Table 1 Precipitation in Germany by Year and Season

Year of precipitation measurement	Amount of precipitation per year, mm or l/m²	Amount of precipitation in winter, mm or l/m²	Amount of precipitation in spring, mm or l/m²	Amount of precipitation in summer, mm or l/m²	Amount of precipitation in autumn, mm or l/m²
2011	732.9	183.4	89.5	310.9	114.1
2012	767.5	254.2	110.5	260.8	168.2
2013	778.7	213.5	201.5	187.1	233.3
2014	727.1	118.2	151.4	275.2	148.0
2015	701.3	183.2	148.6	204.5	203.9
2016	733.1	191.5	166.5	231.7	153.4
2017	858.7	116.4	157.9	307.3	224.6
2018	586.3	198.6	140.6	129.4	92.8
2019	735.0	215.8	182.7	174.6	207.2
2020	704.9	223.4	105.6	228.2	149.0

Source: https://ru-geld.de/2021

As can be seen in Table 1, in the last 10 years, the highest amount of rainfall in Germany was observed in 2017, at 858.7 mm, and the lowest rainfall was observed in 2018, totalling 586.3 mm. On average in Germany, the total rainfall measured for spring 2021 was 175.6 mm. According to the German national weather service, DWD, there was around 100 to 150 mm of rainfall in 24 hours between 14 and 15 July. In a single month, the Rheinbach-Todenfeld weather station (North Rhine-Westphalia) recorded 158 mm followed by Cologne-Stammheim (North Rhine-Westphalia) with 154 mm, Klein-Altendorf (Rhineland-Palatinate) with 147 mm and Kall-Sistig (North Rhine-Westphalia) with 145 mm.

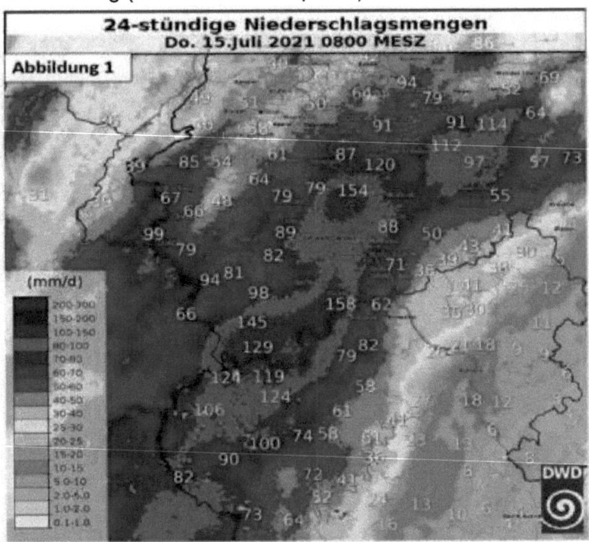

Figure 5 Accumulated 24-hour rainfall figures/Germany 2021.

Source: https://public.wmo.int

Figure 5 shows a greater concentration of precipitation that occurred in the 24-hour period, which is considered extremely high and responsible for the catastrophic flooding that occurred in Germany with great loss of property and life.

Figure 6 Bad Neuenahr-Ahnweiler/2021 Figure 7 https://www.theatlantic.com
Source: https://www.theguardian.com/world/gallery/2021/jul/16/aftermath

Figure 8 Blessen/Erftstadt in July 2021 Figure 9 Kreuzberg on 17 July 2021
Source: https://public.wmo.int/2021 **Source:** https://www.theatlantic.com/

In Figure 6, you can see that a village of houses has been built by the river and on the lower slopes of the mountains. The arrow points to areas that are not only mountainous but also covered in vegetation. Figure 7 shows a mountainous area covered with extensive vegetation. In Figure 8 you can see the wear and tear on the soil caused by the volume of rain that fell in 24 hours in Blessen, in the state of North Rhine-Westphalia in Germany, on Friday. The volume of water was so intense that it opened up a huge fault called a vo^oroca, which led to homes in the area being damaged. In Figure 9, the arrow indicates a mountainous area with a steep slope and overgrown with vegetation, in the image of which we can see a tree uprooted and thrown into the street. This mountainous terrain favours the accumulation of water projected at once in a short space of time, characterising the phenomenon that is the subject of this study, the water header common in sloping areas over which large amounts of rain fall.

Figure 10 shows a large number of tree trunks uprooted when the water, along with debris and mud, comes down from the higher areas, pushing the vegetation along the way and causing destruction to houses located in the lower areas.

Figure 10 Schuld Germany, on July 16, 2021
Source: www.theatlantic.com

Figure 11 Cleaning up water-borne debris
Source: Bad Neuenahr-Ahrweiler, Germany, on July 17, 2021.

Figure 11 shows several houses very close to the railway line and the motorway, all located close to areas with a steeper slope, and this is a characteristic of the headwaters, whose volume drags everything in its path.

Precipitation in Saudi Arabia

Saudi Arabia has an arid and hot desert climate. Its annual rainfall is around 100mm between January and April and in the less rainy months rainfall is around 10mm per month. The number of rainy days averages 45 per year.

Figure 12 The beginning of the rains Figure 13 Camels being dragged/2018

Figure 14 a and 14b Cars and Camels Swept Away by Flash Floods in Saudi Arabia/2018
Source: h ttps://www.you tu be .com/watch ?v=Oh 8PCXKwTKg&t=7 5s

Figure 12 shows an image of the clouds that gave rise to the catastrophic event that struck the city, recorded in Figures 13 and 14 a and b. The speed with which the water descends from the highest regions sweeps away everything in its path, as was the case with the cars that were inadvertently positioned very close to where the current was passing, taking them along with the people. The speed of the water swept away several camels, as shown in Figure 14b.

Figure 16 Formation of the Cabega D'agua. Rainfall in Saudi Arabia equivalent to 6 years in a few hours/2021

Figure 17 Gumayjan Abdullah / Sudden arrival of Cabega D'Agua

Source: https://www.youtube.com/watch?v=7E57dfGlq4E&t=79s The arrows positioned at the top of Figure 17 indicate the mountains, on which a large amount of rain fell in one go, forming a head of water. The arrow in the lower portion of Figure 17 shows a lorry that has moved away from the group shown in the image. It can be seen that the cars are positioned facing the bed on which the water will slide, at which point everyone is taken by surprise by the volume of water and its speed as it descends from the higher ground.

Figures 18 a and b The streets of Mecca have turned into rivers. Saudi Arabia.

Source: https://www.youtube.com/watch?v=CvrW9SQB54k

Even though it has rained a lot, people leave their homes to fulfil their tasks for the day. Then, because they are unaware of the events that could result from the rain, they are taken by surprise, often without even being able to get out of their cars and are swept away by the current, as shown in Figures 18, 19 and 20.

Figure 19 Descent of the Cabega D'Agua F igure 20 Arrival of Water in the Urban Centre
Source: https://www.youtube.com/watch?v=CvrW9SQB54k

Figure 19 shows the outflow of the large volume of water, which is characterised by the head of water that descends at full speed from the highest areas, taking everyone by surprise in urban centres, as can be seen in Figure 20, often causing fatalities.

In order to understand how a waterfall is formed, we watched several videos on YouTube to find an image that would give us an idea of how the phenomenon is formed. Unfortunately, the only one we found was at sea, but even so, we can imagine the damage that a large volume of water causes when it falls on the highest parts of the land at once.

Figure 21 **a** Formation Figure **b** Start of Discharge Figure **c** Precipitation

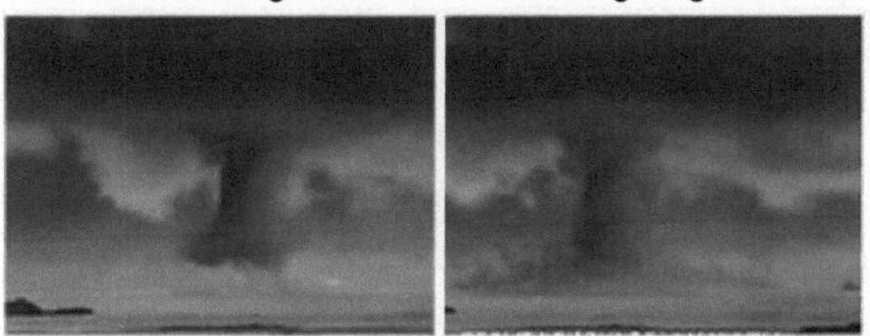

Figure **d** Descent Figure **e** Waterfall
Source: Santos/2022/https://www.youtube.com/shorts/3bXPfdmSrGw

2 Final considerations

The phenomenon we are studying in this research is called Cabega D'agua, which is characterised by a voluminous discharge of rain in a short period of time, usually in areas of high relief, which favours the carrying away of soil and debris from the vegetation, dragging everything in its path. This phenomenon, which is the subject of this study, has occurred a few times, mainly in the south-eastern region of Brazil, due to its higher terrain. However, in the last three decades, various authors have reported a considerable increase in precipitation discharges worldwide. Countries such as Saudi Arabia, whose annual rainfall is often less than 300mm a year, have had rainfall amounts greater than this in less than 48 hours, which characterises an anomaly that has been repeated in several countries at the same time. The causes of these climatic events have yet to be discovered, and claiming one reason or another is not common sense. As the Cabega D'Agua phenomenon is sudden, it can't be avoided in urban centres when it takes people by surprise, as shown in the figures, most of whose images were taken from videos made by residents and passers-by. The best solution for the time being would be to move people away from risk areas, i.e. closer to rivers, even small ones, and away from the base of mountains, in an attempt to avoid disasters such as landslides. The phenomenon under study has been occurring in a number of countries with very varied climates and, as we can see, Figures 21 a, b, c, d, e were taken from a video recorded at sea. What climatic phenomenon over the last two decades has contributed to the formation of water heads in countries like Saudi Arabia, where annual rainfall is very low? Countries such as Belgium, Germany, Sweden and Italy have seen large volumes of water flowing down from higher ground and causing chaos in urban areas. The same can be seen in the higher regions of China, India, Tibet and Japan facing the same challenge. What can we expect in the coming years?

3 References

AZEVEDO, Julia. Understand what Cabega D'agua is and how it happens.
Available at: https://www.ecycle.com.br/cabeca-dagua/ Access: 18:52hs 11/09/2021
Bad Neuenahr-Ahrweiler
https://www.theatlantic.com/photo/2021/07/photos-catastrophic-flooding-across-western-europe/619473/ on 14/09/2021 at 10:57 am.
Cars and Camels Swept Away by Flash Flood in Saudi Arabia/2018
Source: https://www.youtube.com/watch?v=Oh8PCXKwTKg&t=75s 00:06 on 14/09/2021. *Image.*
A truck speeds away from a flood in Saudi Arabia that hit the road at King Abdullah in Houta Bani Tamim, Saudi Arabia. *Image.*
https://www.youtube.com/watch?v=7E57dfGlq4E&t=79s 00:21 on 14/09/2021
CHAVES, Rosane Rodrigues. CAVALCANTI, Iracema Fonseca Albuquerque. Extreme precipitation events over the south of the Northeast. CPTEC/INPE. Pp.1002 -1008. 1985.
COLLISCHONN, Walter. MASATO, Kobiyama. The Hydrology of the Cabega D'agua. Conference:XXIII Brazilian Symposium on Water Resources. Foz do Iguagu, Brazil. December, 2019.
Floods Across Belgium, Germany and *Neetherland/Image*
https://www.theguardian.com/environment/2021/jul/16/climate-scientists-shocked-by- scale-of-floods-in-Belgium On 14/09/2021 at 10:16h
FREITAS, Eduardo de. "As aguas como modeladoras do relevo"; Brasil Escola.
Available at: https://brasilescola.uol.com.br/geografia/as-aguas-como-modeladoras-relevo.htm. Accessed on 16/09/2021 at 00:35hs.
GUMAIJAN, Abdullah - Truck Accelerates to Escape the Water Head
https://www.youtube.com/watch?v=7E57dfGlq4E&t=79s on 14/09/2021 at 11:*08hs/Image*
LEHMANN, lascha. COUMOU, Dim. FRIELER, Katia. Increased record-breaking precipitation events under global warming. Climate Change. 132(4), August 2015.
DOI: 10.1007/s10584-015-1466-3
NBC News - Germany Floods. https://www.nbcnews.com/news/world/climate-change- made-deadly-germany-floods-9-times-more-likely-n1277506 Accessed on 16/09/2021 at 12:14 pm.
NUNES, Aline de Araujo. PINTO, Eber Jose de Andrade. BAPTISTA, Marcio Benedito. Detection of trends for extreme events of precipitation in the Metropolitan Region of Belo Horizonte through statistical methods. Brazilian Journal of Water Resources. Porto Alegre, v.23, e9, 2018.
DOI: http:// dx.doi.org/10.1590/2318-0331.038170134
Public.WMO - 24 Hour Accumulated Precipitation Values/Germany 2021.
Available at: https://public.wmo.int
Public WMO - Heavy rainfall has triggered devastating flooding causing dozens of casualties in Western Europe/ *Image*
https://public.wmo.int/en/media/news/summer-of-extremes-floods-heat-and-fire 09:04h on 14/09/2021
SANTOS, S R Q. BRAGA, C C. SANTOS, A P P, BRITO, J I B. CAMPOS, T L O B. Classification of extreme precipitation events on multiple time scales in Belem-PA. Revista Brasileira de Geografia F^sica, 7, 628-635, 2014.
SANTOS, Anderson. Channel/youtube/ Available at: https://www.youtube.com/shorts/3bXPfdmSrGw. Accessed on 20/03/2022 at 16:09hs.
RU-GELDE.DE - Precipitation by Year and Season. Available at : https://ru-geld.de/en/country/weather-and-climate/precipitation.html#annual Accessed 03:02 on 13/09/2021
TheAtlantic-Schuld Germany. Available at:

https://www.theatiantic.com/photo/2021/07/photos-catastrophic-flooding-across-western-europe/619473/ on 14/09/2021 at 11:00. *Image*
The Guardian: Flood in Germany.
https://www.theguardian.com/environment/2021/jul/16/climate-scientists-shocked-by- scale-of-floods-in-germany, 23:41 on 13/09/2021
Wikipedia/Org - Henan Floods
Available at: https://en.wikipedia.org/wiki/2021_Henan_floods#Zhengzhou
Accessed on: 12/09/2021 at 02:20 am
Wikipedia - 2021 European floods
https://en.wikipedia.org/wiki/2021 European floods
on 12/09/2021 at 01:53h.
Wikipedia. 2021 European Floods. Available at:
https://en.wikipedia.org/wiki/2021 European floods
on 12/09/2021 at 01:53h.
WWA - World Weather Attribution - Heavy rainfall which led to severe flooding in Western Europe made more likely by climate change. Available at:
https://www.worldweatherattribution.org/heavy-rainfall-which-led-to-severe-flooding-in-western -eu rope-made-more-l i kely-by-cl i mate-ch ange/ at 23:31 on 13/09/2021.

CHAPTER 6

IS THERE A LINK BETWEEN GALACTIC WAVES AND EXTREME GLOBAL PRECIPITATION EVENTS?

The effect of electromagnetic waves on the increase in global precipitation is a topic that involves the complex interaction between technology, climate and atmosphere. Although electromagnetic waves, such as those used in telecommunications, radar and microwaves, are widespread in our environment, the scientific evidence on their direct contribution to increasing global precipitation is still limited and largely theoretical. However, we can explore the main areas of interest and theories related to this topic.

1. **Localised Atmospheric Heating:**

- Electromagnetic waves, especially microwaves, have the ability to heat materials with which they interact. In theory, if a sufficient amount of these waves is emitted over a specific area, this could heat the atmosphere locally. Localised heating can create conditions of atmospheric instability, which could, in theory, influence cloud formation and precipitation. However, this influence would be extremely localised and unlikely to have a significant impact on the global climate.

2. **Interference with Particles Suspended in the Atmosphere:**

- Electromagnetic waves can interact with particles suspended in the atmosphere, such as aerosols, dust and water droplets. Some research suggests that these interactions could potentially alter the properties of these particles, such as their size or electric charge, which could, in theory, affect the formation of clouds and the coalescence of raindrops. However, these interactions are highly complex and are unlikely to result in a significant increase in global precipitation.

3. **Influence on the Hydrological Cycle:**

- Some speculative studies have explored the possibility that the proliferation of electromagnetic waves, especially low-frequency ones, could, in theory, influence the hydrological cycle in ways that are not yet fully understood. These theories suggest that the waves could potentially alter patterns of evaporation, condensation and precipitation, but again, the experimental evidence for these hypotheses is limited.

4. **Use of Technology for Climate Change (Geoengineering):**

- Geoengineering involves the use of technologies to manipulate the climate, including precipitation. There are "cloud seeding" experiments that use chemical compounds, such as silver iodide, dispersed by aeroplanes to stimulate precipitation. Another method that began in 2010 was the projection of 22 million tonnes of aluminium-barium oxide into the stratosphere in California, Texas and Florida. This practice has also been authorised in Brazil and several European countries. While these practices do not directly involve electromagnetic waves, there is interest in exploring how electromagnetic waves could be used in a similar way to influence precipitation. However, this area is still at an experimental stage and is the subject of considerable ethical and scientific debate.

5. **Electromagnetic Radiation and Indirect Impact on Precipitation:**

- Electromagnetic radiation from the increased use of communication technologies (such as mobile phone towers, satellites, etc.) can have indirect impacts on the environment. However, the direct effects of these radiations on global precipitation are not yet known to be considered minimal. HAARP, which projects microwaves into the Ionosphere to return to the globe, we don't know what it could cause. The greatest concern associated with electromagnetic radiation is related to its effects on human health and biodiversity, but some studies are trying to investigate its action on climate change or global precipitation.

To better understand the possible interactions between electromagnetic waves and the climate, more research would be needed, especially empirical studies that can validate or refute existing theories. In the meantime, direct climate modification using technology remains an area of interest, but with many unresolved questions and ethical implications to consider.

The primary objective of this research was to study the occurrence of galactic or electromagnetic waves and their possible influence on the increase in global precipitation. Starting in 2010, a model was installed based on the methods used in MIMIC-TC, which retraces microwave images of tropical cyclones on a regular basis in order to observe the behaviour and direction of hurricanes. The images containing galactic waves from 2010 to 2019 were obtained directly from the Index of /real- time/mimic-tpw/ausf/anim to maintain the integrity of the information. With the implementation of the new MIMIC-TPW algorithm, we observed images of transparent galactic waves that pass through the image from right to left and from top to bottom, as well as speckles of various shapes that appear and disappear, not occurring every day. We can't attribute the increase in global precipitation to the action of these waves, mainly because they have only been captured since 2010.

1 INTRODUCTION

Morphed Integrated Microwave Imagery (MIMIC) is a data blending technique that combines disparate bands of microwave observations from orbital satellites in order to create near-perfect animations of the advanced brightness features to obtain images of tropical cyclones from five satellite instruments in low Earth orbit. This model is based on the methods used in MIMIC-TC, which depicts microwave images of tropical cyclones in regular time (WIMMERS & VELDEN, 2007). MIMIC-TPW was originally developed as an experimental near real-time visualisation tool using microwave-derived moisture fields to track the development and progression of cyclone-forming easterlies in the North Atlantic Ocean.

Scientists at the Cooperative Institute for Meteorological Satellite Studies (CIMSS) at the University of Wisconsin-Madison have improved an algorithm they developed almost a decade ago to provide meteorologists with satellite-based total precipitable water (TPW) values from around the world, over the sea and now over land. TPW is the total amount of water vapour contained in a column of moist air, from the Earth's surface to the top of the atmosphere. TPW is an important measure for weather forecasting, especially for predicting and tracking tropical cyclones and other severe conditions in the humid tropics (ANDREOLI et al., 2008).

Wimmers & Velden (2007) explain that the algorithm uses data from seven microwave instruments from the US, NOAA, Japanese and European military satellites, incorporates wind values from the US National Weather Service's global climate model, the Global Forecast System (GFS) and accounts for the movement of water vapour. The Morphed Integrated Microwave Imagery in CIMSS - Total Precipitable Water (MIMIC-TPW) algorithm created in 2007 has become a very useful tool for analysing tropical weather. Two major hurricanes, Harvey in August and Irma in September, hit the United States, bringing attention to the Atlantic as a very active area in recent years. Studying the occurrence of galactic waves and their possible influence on the increase in global precipitation was the primary objective of this research.

2 MATERIAL AND METHOD

To obtain the data, Google and YouTube were used as search engines to obtain academic articles and images. Following methodological rigour, the images containing galactic waves from 2010 to 2019 were obtained directly from the Index of /real-time/mimic-tpw/ausf/anim to maintain the integrity of the information, thus not allowing for falsification. The Morphed Integrated Microwave Imagery in the CIMSS - Total Precipitable Water (MIMIC-TPW) algorithm was used as the base model. Some images posted on the Internet, on the Youtube site, were not found in the Index, so they were not used in this work.

3 RESULTS AND DISCUSSION

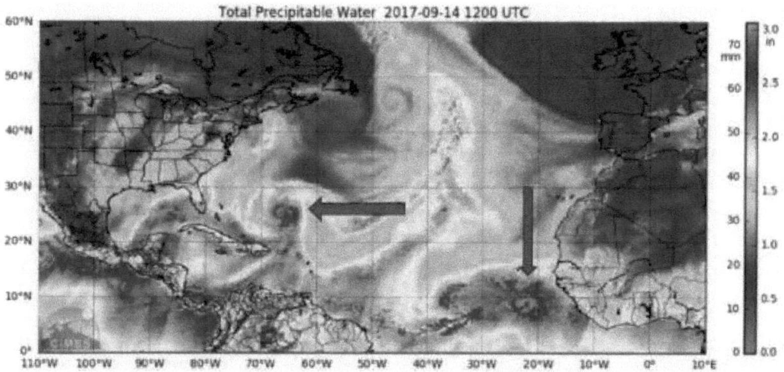

Figure 1 - Total Precipitable Water (TPW) across the Atlantic basin, with highest levels of TPW seen associated with Jose (left) and two tropical waves coming off the African coast (right).

Source: Jet Propulsion Laboratory/California Institute Technology
https://www.jpl.nasa.gov/spaceimages/details.php?id=PIA12096

Total precipitable water is an important factor to consider in the development of a tropical cyclone. The horizontal arrow refers to Hurricane Jose still carrying enough energy to hit the United States. The vertical arrow shows a tropical wave with more energy departing from the Cape Verde Islands showing a stronger signature of precipitable water than Hurricane Jose.

AIRS TOTAL PRECIPITABLE WATER VAPOR (mm), May 2009

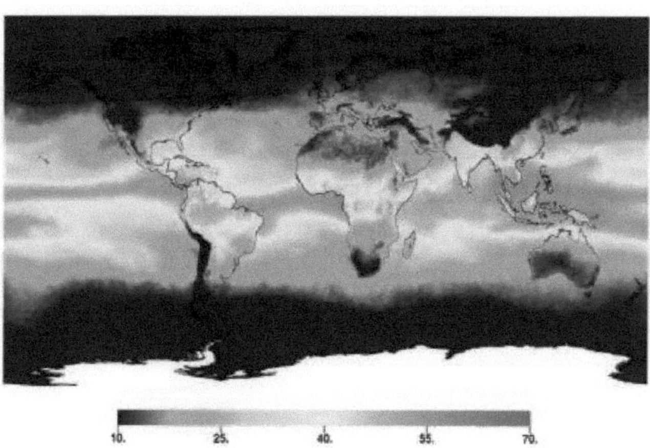

Figure 2 Total Precipitable Water Vapour in May 2009

Source: https://www.jpl.nasa.gov/spaceimages/details.php?id=PIA12096

The image shown in Figure 2 corresponds to the total precipitable water vapour in May 2009 obtained by the Atmosphere Infrared Sounder (AIRS) on NASA's Aqua satellite. The most centralised portion of the vapour column near the equator corresponds to the Intertropical Convergence Zone (ITCZ), a region with strong convection and thunderstorms. The areas furthest from the centre correspond to the cold zones which contain little humidity, unlike the central zone which is very warm and contains more humidity, which condenses to give rise to rainfall.

3.1 Major Storms of Recent Decades

The data in Tables 1 and 2 were obtained from the official website of the CIMSS Tropical Cyclones Page of the Cooperative Institute for Meteorological Satellite Studies (CIMSS), and the data of the recorded events correspond to all the storms that occurred in the Atlantic Ocean between the years 1999/2016.

2016	2015	2014	2013	2012	2011	2010
HI Alex TS Bonnie TS Colin TS Danielle HI Earl TS Я iona H3 Gaston TD EIGHT H1 Hermj ne TS Ian TS Julia TS Karl TS Lisa H5 Matthew H4 Nicole H2 Otto	TS Ana TS Bill TS Claudette H3 Danny TS Erika HI Fred TS Grace TS Henri TD NINE TS Ida H4 Joaquin HI Kate	H1 Arthur TD TWO H1 Bertha H2 CnstobaJ TS Dolly H 3 E doua rd HI Fay H4 Gonzalo TS Hanna	TS Andrea TS Berry TS Chantal TS D oran TS Erin TS Fernand TS Gabrielle TD O8L HI Humberto HI Ingrid TS Jerry TS Karen TS Lorenzo TS Melissa	TS Alberto TS Beryl HI Chris TS Debby HI Ernesto TS Florence H2 Gordon TS Helene HI Isaac TS Joyce H2 Kirk HI Leslie H3 Michael HI Nadine TS Oscar TS Patty HI Rafael H2 Sandy TS Tony	TS Arlene TS Bret TS Cindy TS Don TS Emily TS Franklin TS G ert TS Harvey H3 Irene TD TEN TS Jose H4 Katia TS Lee HI Maria TS N ate H4 Ophelia HI Phillipe H 2 Rina TS Sean	H2 Alex TD TWO TS Bonnie TS Colin TD FIVE H4 Danielle H4 Eart TS Fiona TS Gaston TS Hermine H4 Igor H4 Julia H3 Kart HI Lisa TS Matthew TS Nichole HI Otto H2 Paula H2 Richard HISha.ry H2 Tomas
16	12	□9	14	19	19	21

Adapted from Morphed Integrated Microwave Imagery at CIMSS/Storm Archive
Source: http://tropic.ssec.wisc.edu/real-time/mimic-tc/archive/archive.html

Table 2 Storms that occurred between the years 2010/1999 - Atlantic

2009	2008	2007	2006	2005	2004	2003	1999
TD ONE TS Ana H4 Bill TS Claudette TS Danny TS Erika H3 Fred TD HIOLht TS Grace TS Henri H2 Ida	Arthur Bertha Chris tohaJ. Dolly Edouard Fay Gustav Hanna Ike Josephine Kyte Laura Marco Nana Omar .SwJteea Paloma	Andrea Barry Chantal Dean Erin Felix Gabrielle Ingrid Humberto 1OL Jerry Karen Lorenzo Melissa 15L Noel Olga	Alberto Beryl Chris Debbie Ernesto Florence Gordon Helene Isaac	Arlene Bret Cindy Dennis Emily Franklin Gret Harvey Ire ne Jose Katrina Lee Maria Hate Ophelia Philippe Rita Tammy Vince Wilma Alpha Beta Gamma Delta Epsilon Zeta	Charley Frances Ivan Jeanne	Isabel	Floyd
12	17	17	09	26	04	O1	01

Adapted from Morphed Integrated Microwave Imagery at CIMSS/Storm Archive
Source: http://tropic.ssec.wisc.edu/real-time/mimic-tc/archive/archive.html

The data in Tables 3 and 4 were obtained from the official website of the CIMSS Tropical Cyclones Page of the Cooperative Institute for Meteorological Satellite Studies (CIMSS), and the values of the recorded events correspond to all the storms that occurred in the eastern Pacific Ocean between the years 2004/2016.

2016	2015	2014	2013	2012	2011	2010	2009
TDONE TS Agatha H3 Blas H2 Celia H3 Darby TS Este 1 le HI Frank TS Georgette TS Howard TS Yvette TS Javier TSKay H4 Lester H4 Madeline HI Newton HI Orlene HI Paine TD Roslyn HIUlika H4 Seymour TSTina TS Otto H2P.al.i	H4 Andres H4 Blanca HI Carlos TSEla H4 Dolores TS Enrique TS Eejicja TD EIGHT H2 Guillermo H4 Hilda TD ELEVEN H4 Ignacio H4 Jjmena TS Kevin H3 Linda TD SIXTEEN HI Marty TS Nora H4 Olaf H5 Patricia TS	H4 Amanda TS Boris H4 Cristina TS Douglas TS Elida TS Fausto H5 Genevieve HI Hernan H4 Jselle H3 Mio HI Karina HI Lowell H5 Marie H3 Norbert H4 Odile TD SIXTEEN HI Polo HI Rachel H4 Simon TS Trudy H2 Vance	TS Alvin HI Barbara HI Cosme HI Dalila HI Erick TS Flossie HI Gil H2 Henriette TS Pewa TS Unala TS Ki ко TS Lorena HI Manuel TS Narda TS Octave TS PricjUa H3 Raymond TS Sonia	TS Alena H3 H2 Bud Carlotta H3 Daniel H4 Emilia H2 Fabio HIGi.lma TS Hector HI lleana TS John TS Kristy HI Lane H3 Miriam TS Norman TS Olivia H3 Paul TS Rosa	H4 Adrian HI Beatriz HI Calvin H4 Dora H4 Eugene TS Fernanda HI Greg TD EIGHT H4 Hillary H3 Jova HI Irwin TD TWELVE H4 Kenneth	TS ONE TDTWO TS Blas H5 Celia H3 Darby TDSIX TS Este 1 le TD EIGHT HI Frank TDTEN TD ELEVEN TS Georgette	TDONE HI Andres TS Blanca H2 Carlos TS Dolores TS Lana TS Enrique H4 Felicia TD NINE H3 Guillermo TS Hilda TS Ignacio H4 Jj.rn.ena TS Kevin HI Linda TS M arty TS Nora TS Olaf H5 Rick H3 Neki TS M aka TDTWO- C

	Rick H4 Sandra TS Halola TS lune H4 Kilo HI Loke TS Malia TS Nial.a H2 Oho TD EIGHT-C TD NINE-C	TS Wall HI Ana					H3 Neki
23	29	23	18	17	13	12	"23J-

Adapted from Morphed Integrated Microwave Imagery at CIMSS/Storm Archive
Source: http://tropic.ssec.wisc.edu/real-time/mimic-tc/archive/archive.html

Table 4 Storms between 2004/2008 - East of the Pacific

2008	2007	2006	2005	^004
01W (01W) Neoguri (02W) Rammasun (03W) Matmo (04W) Halong (05W) Nakri (06W) Fengshen (07W) Kalmaegi (08W) Fung-Wong (09W) Kammuri (10W) 11W (11W) Vongfong (12W) Nuri (13W) Fourteen (14W) Sinlaka (15W) Sixteen (16W) Hagupit (18W) Jangmi (19W) MekkhaJa (20W) Higos (21W) Twenty-two (22W) Baui (23W) Maysak (24W) Haishen (25W) Noul (26W) Twenty-seven (27W)	Kong-rey (01W) Yutu (02W) Toraji (03W) Man-yi (04W) Usagi (05W) 06W Pabuk (07W) Wutip (08W) Sepat (09W) Fitow (10W) Danas (11W) Nari (12W) Wipha (13W) 14W Fransisco (15W) Lekima (16W) Krosa (17W) Lingling (18W) Kajiki (19W) Faxai (20W) Peipah (21W) Tapah (22W) Hagibis (23W) Mi tag (24W) 25W (25W) 26W (26W)	01W 02W 03W 04W 05W 06W 07W 08W 09W 10W 11W 12W 13W 14W 15W 16W 17W 18W 19W 20W 21W 22W 23W 24W 25W	01W 02W 03W 04W 05W 06W 07W 08W 09W 10W 11W 12W 13W 14W 15W 16W 17W 18W 19W 20W 21W 22W 23W 24W 25W	19W (Chaba) 26W (Ma-On) 28W (Nock-Ten) 30W (Nanmadol)
27	26	25	25	04

Adapted from Morphed Integrated Microwave Imagery at CIMSS/Storm Archive
Source: http://tropic.ssec.wisc.edu/real-time/mimic-tc/archive/archive.html

The data presented in Tables 1, 2, 3 and 4 suggest that from the end of the 1990s there was a disturbance in global mobile precipitation. In 1999, only one storm case was recorded in the CIMSS storm archive, although 87 cases occurred in that decade up to 2009, as can be seen in Table 2. In the second decade, between 2009 and 2016, 110 cases were recorded, totalling 197 cases in the Atlantic Ocean. The data in Table 3 and 4 together correspond to the number of storms in the Pacific Ocean with a total of 265 storms. The year 2004 registered 04 storms, unlike the others, with the highest number in the second decade with 158 storms.

3.2 Magnetic Energy Waves

Developed as an observation instrument, the Morphed Integrated Microwave Imagery

(MIMIC) was created to provide meteorologists with Total Precipitable Water (TPW) values, as well as to predict and track hurricanes. During its daily operation, it has been possible to pick up waves that are distributed around the globe, but so far the origin of these radiances is unknown. The first validated and scientifically documented wave of so-called galactic energy occurred through the youtuber Dutchsinse, who observed some images forming and then disappearing. We have followed the images year after year since their implementation, from these waves in the hdex of the Morphed Integrated Microwave Imagery at CIMSS, which will be presented below. There have been various theories about these waves, but nothing conclusive to date. When the youtuber mentioned above recorded other energy waves travelling around the world, there were no abnormal weather events in the areas where they appeared. From 2010, when it was first seen on the YouTube website, until the present day, 07 September 2019, few images have been recorded with remarkable clarity. These energy waves, which are registering a very high frequency in the spectrum, do not yet have a defined source.

Since 2010, as the following figures show, the AUS Morphed Composite radar has detected some anomalies that intercept the vapour columns present over the various continents in various directions. The images were obtained by freezing images from the youtube site when the radar is in full activity.

Analysing Images of Galactic Waves

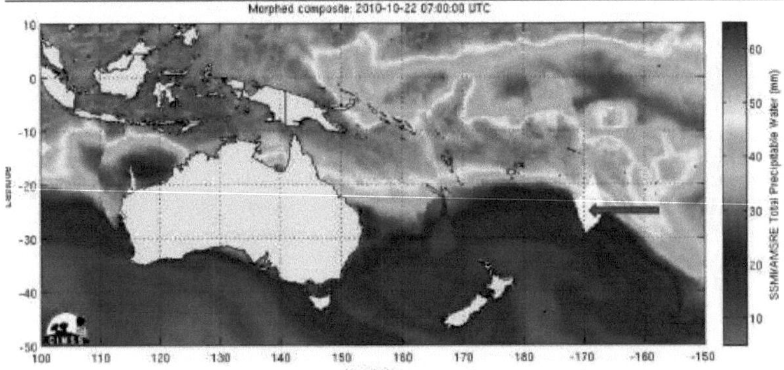

Figure 3 Record of the First Image of the Galactic Wave
Source: Index of /real-time/mimic-tpw/ausf/anim/2010
http://tropic.ssec.wisc.edu/real-time/mimic-tpw/ausf/anim/2010/

The arrow on the right-hand side of the image shows the end of the projection, which is very quick to record. It is important to note that this phenomenon began on 22 October 2010, as shown by the Morphed Composite data located above the image.

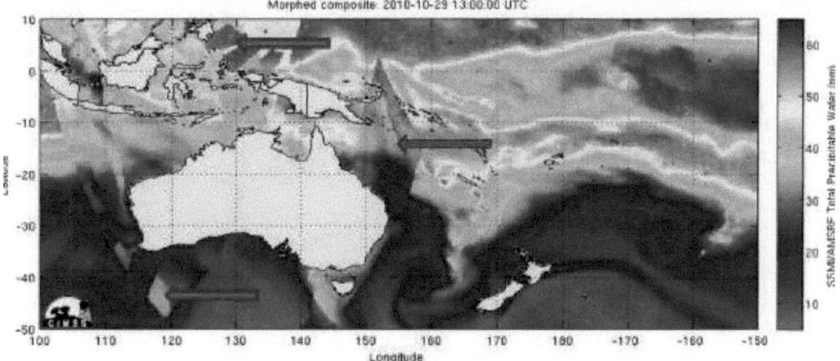

Figure 4 Set of diffuse images of Australia

Source: Index of /real-time/mimic-tpw/ausf/anim/2010

http://tropic.ssec.wisc.edu/real-time/mimic-tpw/ausf/anim/2010

Between 29 and 31 October 2010, some images were taken in which the galactic wave appears to have very distinct shapes that pass to the east, west and north of Australia, as shown by the three arrows indicated in the image.

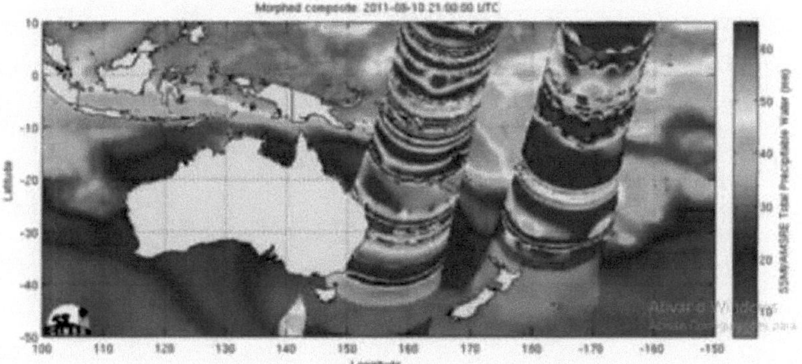

Figure 5 Column-shaped Galactic waves

Source: Index of /real-time/mimic-tpw/ausf/anim/2011

http://tropic.ssec.wisc.edu/real-time/mimic-tpw/ausf/anim/2011

Of all the images taken of the hdex, these two column-shaped ones are considered the most defined in terms of visibility. It's important to note that this phenomenon occurred to the east of Australia.

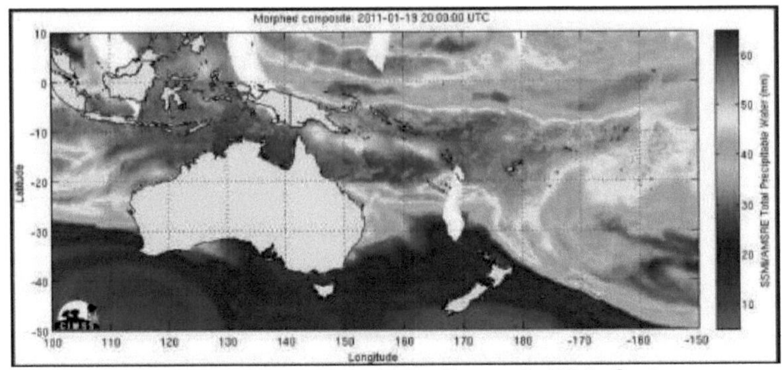

Figure 6 Galactic Waves in the Shape of a White Spot
Source: Index of /real-time/mimic-tpw/ausf/anim/2011
http://tropic.ssec.wisc.edu/real-time/mimic-tpw/ausf/anim/2011
22:50 on 07-09-2019

In Figure 6 you can see the white spots that descend from the upper face and to the east of the image in the direction of the lower face reach New Guinea and Borneo.

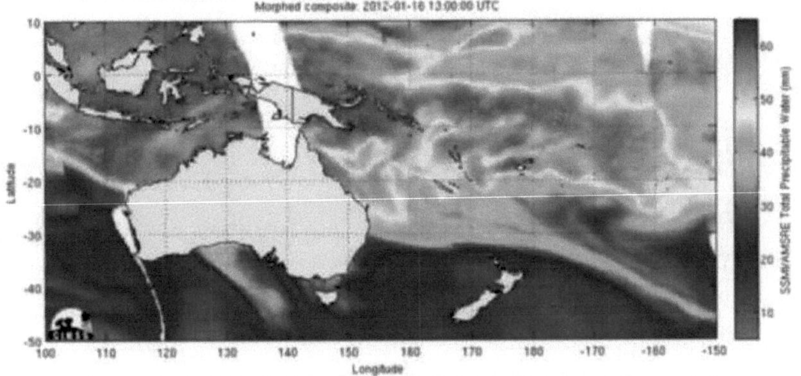

Figures 7 Galactic Waves in the Form of a Partial White Spot
Source: Index of /real-time/mimic-tpw/ausf/anim/2012
http://tropic.ssec.wisc.edu/real-time/mimic-tpw/ausf/anim/2012

You can see several small white shapes in Figure 6, which are repeated in Figure 7, positioned directly over northern Australia. At times, these spots disappear when they reach the country or any area of the ocean and others, which are more distinct, cross the globe always in the direction from the Arctic to the Atlantic.

Antarctica. At no point did we find a reverse pattern in the images.

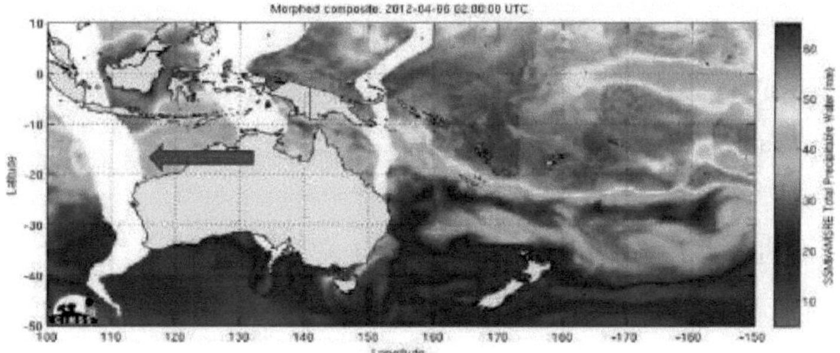

Figure 7.1 Galactic Waves in the Shape of a Total White Spot

Source: Index of /real-time/mimic-tpw/ausf/anim/2012

http://tropic.ssec.wisc.edu/real-time/mimic-tpw/ausf/anim/2012/

Although these are just details seen in the image, it's important to remember that these waves don't follow a single pattern, sometimes appearing as distinct bands, sometimes as transparent stripes projected onto the ocean, not very viscous, like the band indicated by the arrow in the upper right-hand side of the image, while in the upper left-hand side, they appear as white spots that affect the whole of Australia.

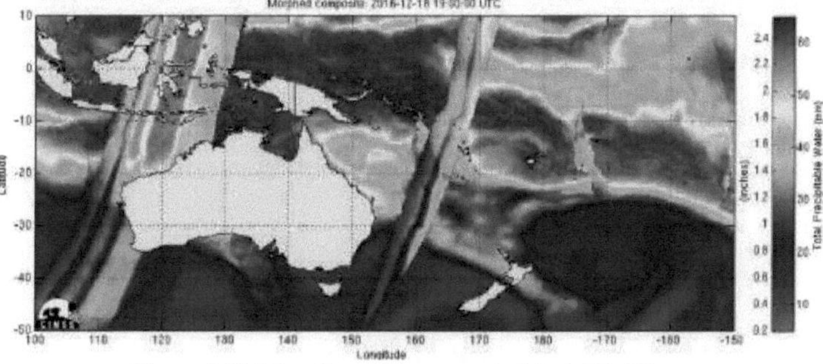

Figure 8 Unidirectional Movement of Galactic Waves

Source: Index of /real-time/mimic-tpw/ausf/anim/2016

http://tropic.ssec.wisc.edu/real-time/mimic-tpw/ausf/anim/2016

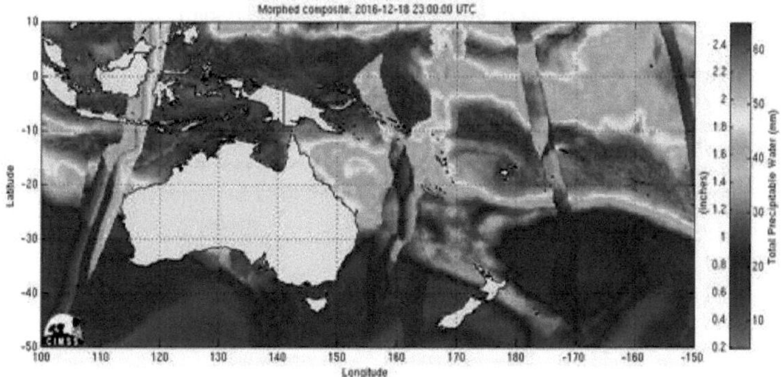

Figure 9 Multidirectional Movement of Galactic Waves

Source: Index of /real-time/mimic-tpw/ausf/anim/2016
http://tropic.ssec.wisc.edu/real-time/mimic-tpw/ausf/anim/2016

Figures 8 and 9 were taken at the same time, with seconds between them. Despite the same date of acquisition of the image located on the upper side, it can be seen that the time recorded in UTC is different, with Figure 8 at 19:00:00 and Figure 9 at 23:00:00. What is striking are the very different directions of these waves taken in 2016.

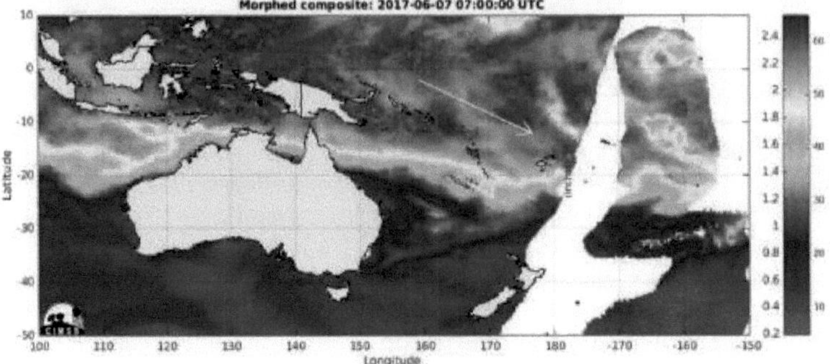

Figure 10 Incomplete White Spots of Galactic Waves Source: Index of /real-time/mimic-tpw/ausf/anim/2017

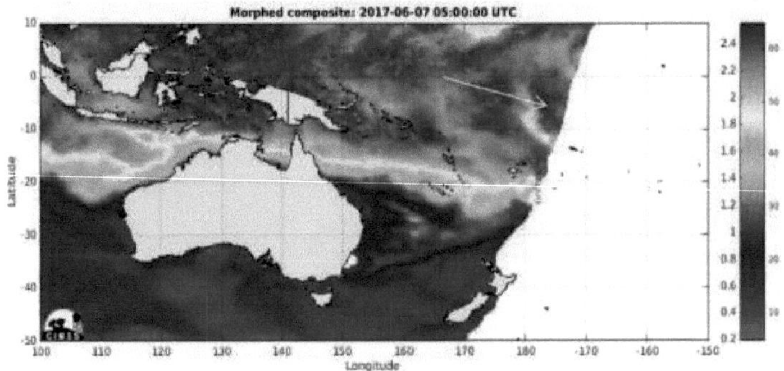

Figure 10.1 Complete White Spots of Galactic Waves
Source: Index of /real-time/mimic-tpw/ausf/anim/2017
http://tropic.ssec.wisc.edu/real-time/mimic-tpw/ausf/anim/2017

The arrows show an abnormal pattern of undetected white spots between the years 2010 and 2016. The two moments recorded in Figure 10 are 7:00:00 and 7:05:00 respectively, repeated in Figure 10.2 only on the opposite side of the image. The radar in the months of January to May 2017 did not show any characteristic images, however, when viewing the animation it is possible to notice the displacement from right to left on all days, of transparent images that move as if they were waves and, so far, there is no explanation for this phenomenon. The other days showed normal images.

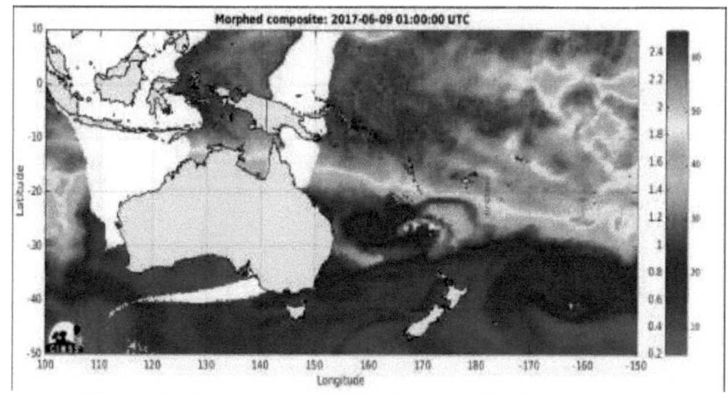

Figure 10.2 Complete White Spots of Galactic Waves
Source: Index of /real-time/mimic-tpw/ausf/anim/2017
http://tropic.ssec.wisc.edu/real-time/mimic-tpw/ausf/anim/2017

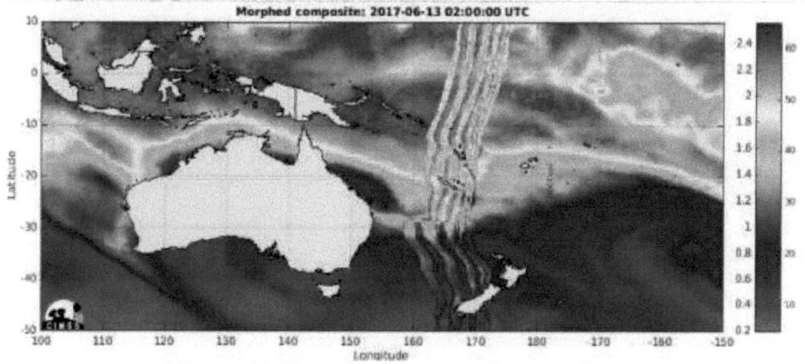

Figure 11 Galactic spots arranged in ribbons
Source: Index of /real-time/mimic-tpw/ausf/anim/2017
http://tropic.ssec.wisc.edu/real-time/mimic-tpw/ausf/anim/2017

In the image in Figure 11, the presentation of galactic waves in the form of parallel ribbons differs completely from the patterns observed previously.

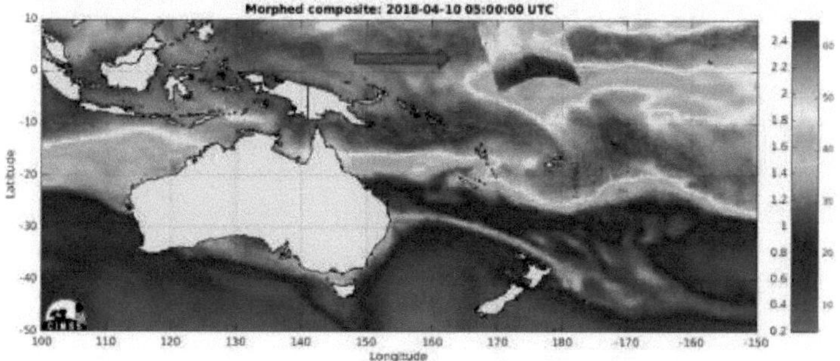

Figure 11 Galactic Spot Arranged in a Column
Source: Index of /real-time/mimic-tpw/ausf/anim/2018
http://tropic.ssec.wisc.edu/real-time/mimic-tpw/ausf/anim/2018

The 2018 records were normal and the images were unchanged, but it was possible to see the image in the shape of a cylinder, as shown by the arrow, for a few days.

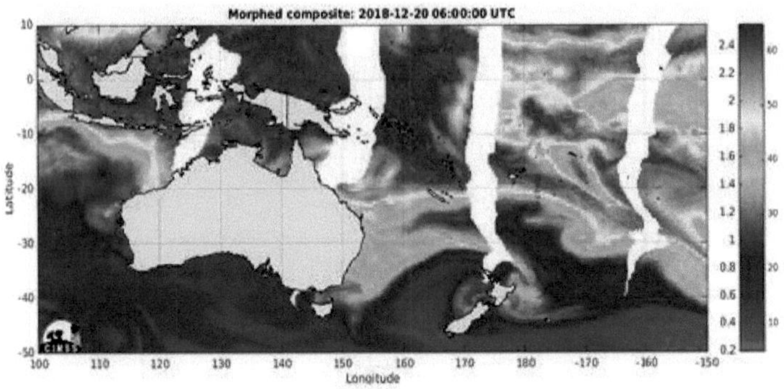

Figure 12 Galactic spots in bands
Source: Index of /real-time/mimic-tpw/ausf/anim/2018
http://tropic.ssec.wisc.edu/real-time/mimic-tpw/ausf/anim/2018

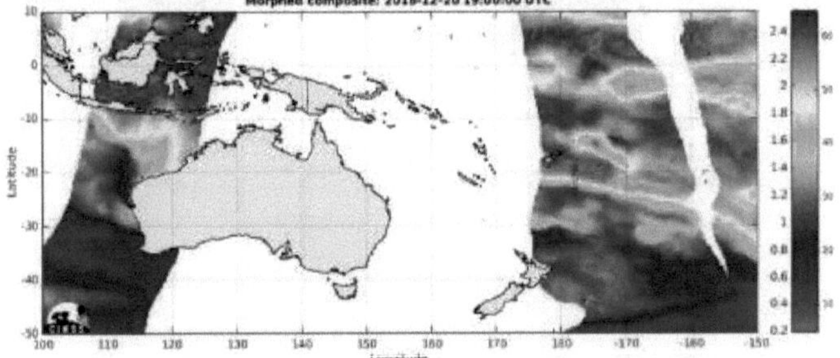

Figure 12.1 Abnormal Pattern of Galactic Spots
Source: Index of /real-time/mimic-tpw/ausf/anim/2018
http://tropic.ssec.wisc.edu/real-time/mimic-tpw/ausf/anim/2018

Between January and November 2018, we didn't see any significant changes in the images provided by the Morphed Composite, but in December it was possible to obtain images as shown in Figures 12 and 12.1, with the latter showing very visible white spots, almost blotting out the image.

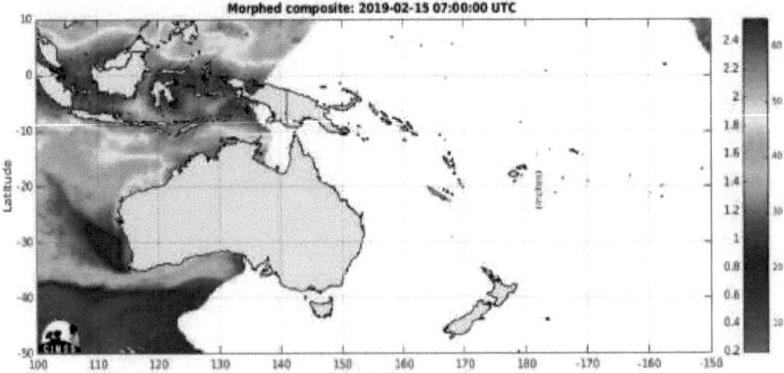

Figure 13 Abnormal Pattern of Galactic Spots
Source: Index of /real-time/mimic-tpw/ausf/anim/2019
http://tropic.ssec.wisc.edu/real-time/mimic-tpw/ausf/anim/2019

3.3 Increase in global precipitation

Precipitation is considered to be one of the most important processes for the Earth's biosphere, including humanity. For the oceans, precipitation is a source of fresh water and therefore an important forcing term for the thermohaline circulation. In the atmosphere, precipitation releases latent energy, which is the main forcing term for atmospheric movements, especially in the tropics. The average global rainfall is equivalent to a column of water about 1 metre high. This is equivalent to a continuous supply of energy to the atmosphere of around 80 Wm-2, which represents approximately 30 per cent of the solar energy received. Caused mainly by its strong coupling to the vertical velocity field, precipitation is extremely variable in space and time (GLOERSEN & BARATH, 1977).

In most areas of the globe, especially at sea, there are not enough direct measurements to meet the needs of a proper understanding of the hydrological cycle. In addition, measurements on land are hampered by orographic effects. At sea, errors are caused by the obstruction of the airflow caused by the ship. Remote sensing from the surface by radar allows for high temporal sampling and greater area coverage, but cannot be applied globally due to the high costs of equipment and personnel. Furthermore, the inversion of radar measurements is not direct and presents its own problems, such as the uncertainty of the falling size spectra and the increase in the elevation of the radar beam from the earth's surface with increasing distance. Remote sensing from 124 satellites is left as the only practical method for achieving global coverage of precipitation estimates (GLOERSEN & BARATH, 1977).

Precipitation plays a vital role in understanding the mechanism and interaction of the global water and energy balance and is therefore the main provider of data for the creation of climate models and studies (HONG et al., 2019).

The development of remote sensing and satellite techniques has increased human capacity to observe the phenomenon of precipitation on a global scale (CDES, 2018, p.265).

Researchers from various countries have come to the same conclusions about the increase in precipitation, which has recently been causing economic and social damage, including a large number of victims. Ruezzene et al. (2017) observed an upward trend in temperature and rainfall in most seasons. To analyse precipitation in Italy, Buffoni et al. (1999) analysed data from 32 stations over a period of 164 years, verifying the change in volume over the last few decades. Rocha (2011) observed an increase in precipitation in the city of Recife between the years 1990 and 2009, in the last decade whose value was double the values observed in the first. In analysing monthly precipitation, Domonkos (2003) observed an increase in precipitation trends between 1901 and 1998 and its possible relationship with atmospheric changes in Europe and the Atlantic Ocean. Bodri et al. (2005) studied the trend in precipitation changes in the Czech Republic between 1994 and 2001 and their results show a significant increase in the variability of precipitation in this country. Saeidabadi et al. (2016) studied the climatology of Ira and observed that precipitation increased considerably between the years. When studying precipitation in Russia, Krokhin & Luxemburg (2007) observed a sharp increase in precipitation intensity. Dordevic (2008) observed an increase in maximum daily and monthly rainfall for the city of Belgrade. Nguyen et al., (2018) recorded a significant increase in precipitation on a global scale.

4 FINAL CONSIDERATIONS

The MIMIC-TPW algorithm that captured the images of these waves began in 2010, but we don't know if they have been occurring throughout the human era, and they have been more potent in the last two decades. It's difficult to attribute the increase in global precipitation to any man-made process, especially since environmental disasters already exist on planet Earth, so we can't attribute the responsibility for the increase in global precipitation to these processes either, especially since the latest records of extreme mobile precipitation have occurred in specific countries and locations, such as the Saudi Arabian desert and neighbouring countries, which, between the months of November and December, received in just a few days a rainfall expected for several years. It is important to remember that between the years 2010/2021, various abnormal events such as locust invasions, hailstorms, extreme temperature rises, waves of snow coming off the sea and invading homes, frozen rivers in European countries, river droughts and fires in California, changes in the behaviour of the seas, invading in some places and retreating in others, have occurred in several countries on several continents. To date, scientists have not reached a conclusion about the origin of these magnetic waves, which have different colours and shapes. It's important to note that the waves always travel in the same direction, i.e. from the top to the bottom of the map.

5 REFERENCES

Andreoli, Rita V. Ferreira, S H S. Sapucci, Luiz F. Souza, Rodrigo A F. Mendonca, R W B. Herdies, Dirceu L. Aravequia, Jose A. Contribution of different observation systems to weather forecasting at CPTEC/INPE. 2008. Revista Brasileira de Meteorologia, v.23, n.2, pp.219-238.

Bodri, L. Cermak, V. KRESL, M. Climate Change. 72:151-170. 2005. https://doi.org/10.1007/s10584-005-5370-0

Buffoni, L. Maugeri, M. Nanni, T. Precipitation in Italy from 1833 to 1996. Theor Appl Climatol. 63, 33-40.1999. DOI: 10.1007/s007040050089

CDSES-Committee on the Decadal Survey for Earth Science. Thriving on Our Changing Planet: A Decadal Strategy for Earth Observation from Space. 716p. Ed. The National Academies Press. Washington. 2018. DOI:10.17226/24938.

Dordevic, Smilja V. Temperature and Precipitation Trends in Belgrade and Indicators of Changing Extremes for Serbia. Geographica Pannonica. Vol.12, Issue 2, pp.62-68. 2008.

Domonkos, P. Recent precipitation trends in Hungary in the context of large scale climatic changes. Natural Hazards. 29. Pp.255-271. 2003. DOI: 10.1023/A:1023690014955

Index of /real-time/mimic-tpw/ausf/anim. Available at: http://tropic.ssec.wisc.edu/real-time/mimic-tpw/ausf/anim/?M=D Accessed on 09/09/2019 at 12:05 pm.

Jet Propulsion Laboratory/California Institute of Technology. Global Total Precipitable Water Vapor for May 2009. AIRS Atmospheric Infrared Sounder Available at: https://www.jpl.nasa.gov/spaceimages/details.php?id=PIA12096 accessed on 09-09-2019 at 19:27hs.

Krokhin, V V. Luxemburg, W M J. Temperatures and precipitation totals over the Russian Far Eastand Eastern Siberia: long term variability and its links toteleconnection indices. Hydrol. Earth Syst. Sci., 11, 1831-1841. 2007. DOI: https://doi.org/10.5194/hess-11 -1831 - 2007

Nguyen, Phu. Thorstensen, Andrea. Sorooshian, Hsu et al. Global Precipitation Trend across spatial Scales Using Satellite Observations. 10p. American Meteorological Society. BAMS, april.2018. DOI https://doi.org/10.1175/BAMS-D-17-0065.1

Rocha, Joao Vitaliano de Carvalho. Santos, Valdemir Alexandre. Characterisation of Climate Change in the City of Recife Using Time Series. Ano 13-vol.20- Jan/jul. 2017. DOI: http://dx.doi.org/10.5380/abclima.v20i0.43518

Ruezzene, Bermond. Fambri, Larissa Santos. Aguiar, Renata Gongalves et al. Climatic Trends in Time Series of Maximum Temperature in Porto Velho-RO. 6p. X Brazilian Workshop on Micrometeorology, Santa Maria-RS, 8-10 November 2017.
Available at: https://even3.blob.core.windows.net/anais/63862.pdf Accessed on 09/09/2019 at 15:25hs.

Saeidabadi, R., Najafi, M.S., Roshan, G. et al. Asia-Pacific J Atmos Sci . 52: 437. 2016. https://doi.org/10.1007/s13143-016-0026-8

Wimmers, A J. Velden, C S. Seamless Advective Blending of Total Precipitable Water Retrievals from Polar-Orbiting Satellites. 2011. J.Appl.Meteor. Climatol., 50, 1024-103. doi: http://dx.doi.org/10.1175/2010JAMC2589.1

CHAPTER 7

ANALYSING THE ENVIRONMENTAL IMPACTS OF WILDFIRES IN THE STATE OF CALIFORNIA FROM 1999 TO 2017

Forest and urban fires have significant and varied environmental impacts, which can affect the ecosystem, the atmosphere, biodiversity and even human health. These impacts are often devastating and can have long-term consequences. Below are the main environmental impacts caused by fires:

1. Loss of Biodiversity:

- **Habitat destruction**: Forest fires can destroy vast areas of natural habitat, resulting in the loss of flora and fauna. Animal species that cannot escape in time can be killed, and plants, especially those that are not adaptable to fire, can be completely destroyed.
- **Reduction of Genetic Diversity**: The destruction of local populations can reduce the genetic diversity of affected species, making them more vulnerable to diseases and other threats.

2. Soil Erosion and Degradation:

- **Erosion**: After a fire, the vegetation that normally holds the soil with its roots is destroyed, leaving the soil exposed to erosion. This can result in landslides, especially in mountainous areas, and significant loss of fertile soil.
- **Reduced Fertility**: Fire can destroy the nutrient-rich topsoil, leading to a decrease in fertility and affecting the ability of vegetation to regenerate.

3. Gas and Pollutant Emissions:

- **Release of Greenhouse Gases**: Forest fires release large quantities of carbon dioxide (CO_2) and other greenhouse gases, such as methane (CH_4), which contribute to global warming.
- **Air Pollution**: Smoke from fires contains fine particulates (PM2.5), carbon monoxide (CO), nitrogen oxides (NOx), and volatile organic compounds (VOCs), which can cause serious respiratory and cardiovascular problems for both humans and animals.

4. Contamination of Water Resources:

- **Water Pollution**: Ash and residues from fires can be carried into rivers and lakes by rainfall, contaminating drinking water sources with heavy metals and other toxic pollutants.
- **Disruption of Hydrological Cycles**: The destruction of vegetation alters the water cycle, reducing the soil's ability to absorb rainwater, which can lead to flooding and reduced water replenishment.

5. Impact on Vegetation and Forest Regeneration:

- **Changes in vegetation composition**: Some plants may be more resistant to fire, while others are completely eliminated. This can result in changes in the composition of plant species in an area, favouring invasive or less desirable species.
- **Difficulty in Regeneration**: In heavily affected areas, the natural regeneration of vegetation can be extremely slow or impossible without human intervention.

6. Local Climate Change:

- **Microclimate change**: The loss of vegetation can alter the local microclimate, affecting temperature, humidity and wind patterns. This can make it even more difficult for the ecosystem to recover.
- **Heat Island Effect**: In urban areas, fires can exacerbate the heat island effect, where the affected areas become significantly hotter than the surrounding areas.

7. Impact on Human Health and Quality of Life:

- **Health Problems**: Inhaling smoke can cause respiratory problems such as asthma and bronchitis, as well as increasing the risk of cardiovascular disease. People with pre-existing conditions are particularly vulnerable.

- **Displacement of Communities**: Large-scale fires can force the displacement of populations, resulting in significant social and economic impacts.

8. **Interruption of Biogeochemical Cycles:**

- **Carbon Cycle**: Fires release carbon stored in biomass into the atmosphere, disrupting the carbon cycle and contributing to an increase in CO $._2$
- **Nutrient Cycle**: The destruction of vegetation can disrupt the cycle of nutrients such as nitrogen and phosphorus, which are essential for soil health and plant growth.

Mitigation Measures:

- **Prevention and control**: Fire prevention programmes, such as controlled forest management, the creation of firebreaks and awareness campaigns, are essential to minimise the risks.
- **Ecosystem restoration**: After a fire, reforestation and habitat recovery efforts are needed to restore biodiversity and ecosystem functionality.
- **Improved Environmental Policies**: Implementation of stricter policies to reduce the human causes of fires, such as uncontrolled agricultural burning and inadequate waste management. Wildfires have devastating environmental consequences that go beyond the immediate destruction of forests and fields. The recovery of these ecosystems can take years or even decades, and human actions play a crucial role in preventing and mitigating these impacts. Studying the occurrence of the multiple wildfires that have taken place over the last decade in the state of California was the primary objective of this research. California is a region in which the largest industries are concentrated. It also has natural recreation and conservation areas, a military, a natural wildlife reservoir, co-operative plans for the conservation and development of fish and wildlife administered by the Navy. Google was used as a search engine to obtain data through articles and websites, as well as images obtained through remote sensing. It was observed that the large number of fires in California has not yet been justified by scientists or by the agendas responsible for suppression. The large number of wildfires in California has caused unprecedented destruction in recent decades, affecting homes and businesses, agricultural, livestock, leisure and recreational areas, completely de-characterising the region, despite the fact that logging is so common.

1 Introduction

The impacts of forest fires on the environment are so complex and multi-faceted that a complete analysis will only be possible if we draw on a wide range of sciences, including the different specialities of geography, which must play a leading role in detecting the origin of the cause. In order to do this, the analysis must focus not only on the immediate environmental impacts, but also on the impacts following the fire. The multiple, complex and varied environmental impacts depend essentially on various factors linked to the characteristics of the fire itself, such as its size, the intensity of the fire, the time of year, location, type of vegetation, duration and frequency of fires.

Other factors that condition the development of forest fires have to do with fuels (volumes available, concentration, characteristics and how they are distributed) which, in turn, depend on the properties of the composition and size of forest stands, often resulting from the nature and characteristics of the soil itself (Guimaraes, de Souza, Fiedler, & da Silva, 2014). Wildfire activity in the western United States, more particularly in California, has been on the rise in recent years, causing major economic impact and, above all, extensive destruction with loss of life (Lourengo, 1992).

Fire activity has been spreading through the undergrowth to areas with large trees, degrading vast tracts of land. Westerling et al. (2006) believe that the increase in wildfires in the western region of the USA is related to the high temperatures during the spring and summer seasons and the low rainfall, which makes the vegetation drier for a long time. It is likely that the incidence of forest fires is strongly associated with the high temperatures that occur in the dry seasons, when humidity and rainfall are considered very low. Although this scenario is a simplification of two extremes among the wide variety of vegetation types and fire regimes found in California, they provide a useful context for understanding the statistical relationships between fire activity observed historically and the variations that occur in climatological phenomena (Westerling, 2006).

Analysing the impacts generated by the action of a forest fire, Guimaraes et al. (2014) found the following changes: in the human environment, soil heating, erosion, reduction of soil nutrients, reduction of organic matter content, soil acidity, alteration of the local microclimate, reduction of water quality, release of CO2 into the atmosphere; in the biotic environment, animal deaths, wood damage, clearing of the understorey, emission of sprouts, impact on agricultural crops, interference in plant succession, reduction in the activity of micro-organisms and pasture renewal; in the socio-economic environment, inconvenience to the surrounding population, landscape impacts and destruction of recreational areas. The main objective of this research was to study the occurrence of the multiple fires that have taken place in California over the last decade and their possible causes.

1.1 History, land use and zoning control in California

The coastal zone or continental border region is defined here as the area of the 149 Environmental Impacts of Wildfires, California high tide inland 1,000 metres or approximately half a mile. For the purposes of describing land use and zoning in the area of contact between the ocean and the landmass, this definition is adequate, although other purposes may have a more general perspective for a more broadly defined coastal zone. Land use patterns in the coastal zone have their origins in the early 1800s, when Spanish-speaking missionaries and settlers obtained large plots of land and pursued agricultural and pastoral economies.

Dailey, Reish, & Anderson (1993) assert that the post-statehood era (1850) witnessed the subdivision of the largest, most accessible land holdings, widespread land speculation, often followed by city financing and embryonic urban development in favoured locations along the coast. Where the configuration of the coast was suitable, the influx of World War II port population sought coastal locations for recreation and/or domestic use. Federal and state governments own parcels of land in the coastal border area. The development and use of land in state or federal areas is determined by the federal or specific state agency that receives jurisdictional control. The development and use of federally owned land is subject to the provisions of the National Protection Agency.

Environmental, while development and land use on public and private lands are subject to the provisions of the California Environmental Quality Act of 1970.

1.2 California, environmental protection area

Southern California's coastal zone displays a variety of wildlife, ecological and marine reserves, as well as research, integrated management and conservation stations (Galik & Grala, 2017). These features are located on the mainland, in the intertidal zone, in the sea and in some of the channels that surround the islands. Dailey et al. (1993) considers ecological reserves and marine life refuges and reserves to be areas of land and/or water established in accordance with the law by the state of California for the purpose of protecting rare or endangered wildlife or aquatic organisms or specialised terrestrial and aquatic habitat types. Such reserves and refuges are administered by the California Department of Fish and Game. Ecological reserves must be preserved in natural conditions, for the benefit of the general public, to observe native flora and fauna and for scientific study (Horton, Knight, Galvin, Goldstein & Herrington, 2017).

Except in accordance with the regulations of the Fish and Game Commission, it is illegal to enter any ecological reserve or take into it any bird or nest or eggs, or any mammal, fish, mollusc, crustacean, animal, reptile or any form of vegetation or animal life (Christensen et al. 1996). Segerson (2013) considers that one of the greatest difficulties in environmental conservation is the concern of companies not with penalties, but with adverse publicity that could generate informal sanctions generated by the public, the media and, perhaps, the market. The need to improve coastal zone management stems from a strong social requirement to control the behaviour of individuals and organisations that affect and exploit coastal marine ecosystems and their goods and services. Table 1 shows five regions that have become preservation zones due to the extensive areas that have been severely altered by human activities since the 1970s.

Observing the need to preserve the coastal environment, federal, state and local agencies have established areas, as shown in Table 2, between the ocean and the continent to study the relationships that occur between the effects of human activities on terrestrial, estuarine, coastal and oceanic ecosystems with the aim of examining the effects and changes in land use, as well as human impacts on the transport of nutrients, sediments and toxic elements responsible for modifying the landscape in the coastal zone. Military conservation and wildlife management areas. Military installations or managed areas in the coastal zone of southern California have also carried out a wildlife conservation and management programme, in cooperation with the California Department of Fish and Game and the Department of Fish and Wildlife of the US Department of the Interior. The 700-hectare national wildlife reservoir was established in 1972.

In co-operation with the naval authorities, the US Department of Fish and Wildlife manages the wildlife refuge for the conservation and management of migratory birds and other fish and wildlife. Dailey et al. (1993) adds that there is a second programme in which the US Navy, the US Fish and Wildlife agency and the California Department of Fish and Game have developed cooperative plans for wildlife conservation and development on various Navy-managed lands. The co-operative plan calls for species identification and species inventory.

Table I

Ecological reserve in the coastal zone of southern California. Source: Adapted from Dailey et al. (1974)

Table 1

Ecological reserve in the southern California coastal zone. Source: Adapted from Dailey et al. (1974)

Ecological Reserve	Local tza^-io	Remarks
Farnsworth Bank	Los Angeles Co.	The bank is made up of irregu lar rocky pinnacles rising about 50 feet from the surface and contains rare specimens of purple coral Allopora call form ca
Chica bag	Orange Co.	563 hectares bordering the Pacific Coast Highway
Heisler Park	Orange Co.	Adjacent to the town of Praia da Laguna
Buena Vista Logoon	San Diego Co.	71 acres between municipal boundaries of Oceanoside and Carlsbad Oceanside in the
San Diego-La Jolla	San Diego Co.	mare media and alia, north and east of the goldfish points to the latitude of the Scripps Institute of oceanography

Table 2

Heserxcts and *refuges of morning life in Southern California - Content: Adapted from Dailey el al. (1974)*

Table 2

Marine life reserves and refuges in southern California. Source: Adapted from Dailey et al. (1974)

Refugios da Vida Mannha and Bookings	Localisation	Reach of the Sea in the High and Middle Seas	Ocean Frontier (Miles)
Point Fermin	Los Angeles Co.	600	1.00
Newport Beach	Orange Co.	200	o/гз
Irvine Beach	Orange Co.	600	3,41
Laguna Beach	Orange Co.	600	1,45
South Luguna Beach	Orange Co.	600 a -20 Contomo	0.52
Niguel	Orange Co.	1200	2,14
Dana Point	Orange Co.	1200	0.66
Doheny Beach	Orange Co	600	1,1 1
San Diego	San Diego Co.	1OOO	1.00

National monuments

Two national monuments, Cabrillo National Monument and Channel Islands are located in the coastal zone of Southern California. Both monuments are managed by the US National Park Service and, like all national monuments, are designed to protect the locally unique natural environment. Table 3 shows some of the most visited monuments in the region under study, and their preservation is a constant concern for all governments due to the thousands of fires that break out year after year without being controlled or even reduced. Fortunately, many of these parks are located in mountainous areas with less dense vegetation, making it difficult for fires to spread.

Main archaeological sites

A large percentage of the known archaeological rivers in California are located within the coastal zone of Southern California. If all possible categories of rivers (excavated, unexplored, published, unpublished, probable) were counted, the number would be in the hundreds. It is not the intention of this study to include a complete overview of rivers, but rather to record the riches that exist in the various regions that make up the State of California. The areas located between the two mountainous zones and the coastal zone encourage the allocation of agricultural holdings as well as residential uses. In addition, Los Angeles contains extensive low-lying terrain that provides

views of local importance in many of the coastal area's intensely developed cities, including Long Beach, Los Angeles, Torrance, Redondo Beach, Manhattan Beach and Santa Monica. The Channel Islands lie off the coast of Southern California and are under various land uses and jurisdictions. San Miguel (Santa Barbara County), San Clemente (Los Angeles County) and San Nicolas (Ventura County) are administered by the United States Navy and are regularly used for military purposes and are not open to public visits. The US Navy, in cooperation with the California Department of Fish and Game and the US Department of Fish and Wildlife, has established programmes for the conservation and development of fish and wildlife on the islands. Santa Barbara (Los Angeles County) and Ana Capa Islands (Ventura County) were grouped together in 1938 as the Channel Islands National Monument under the jurisdiction of the United States National Park Service. Visitors are few and the preservation of the natural habitat is the main objective of the Monument's management. Material and method The methodological procedure employed to carry out this review used the Google search engine, which searched for national and international articles from 1972 to 2017 in the databases of Elsevier Publishing, SciELO - Scientific Electronic Library Online, Medline and the US National Library of Medicine National Institute of Health, books, blogs and dissertations. As inclusion and exclusion criteria for this study, articles were selected from electronic databases whose subjects objectively addressed the triggers of wildfires in the State of California and their comorbidities at a regional level, as well as information from other databases that were related to the topic, respectively.

Characterisation of the Study Area

The State of California is made up of 27 counties and a variety of regions with very distinct geomorphological features, and areas that experience sub-zero temperatures and fluctuations often exceeding 50°F. It is located at a latitude of 36°77'83" and a longitude of 119°41'79". Climatic considerations about the state of California California's climate varies greatly from arid to subarctic depending on latitude, elevation and proximity to the coast, with the coastal and southern parts having a Mediterranean climate with very wet winters and dry summers. The influence of the ocean does not allow temperatures to be extreme, thus creating warmer winters and relatively cool summers, especially in coastal areas. High-relief areas can experience snowstorms during the winter months. Eastern California is prone to summer storms because of the North American monsoon, but the dry weather during the rest of the year creates favourable conditions for forest fires. Heavy rains can cause floods that favour landslides from steep slopes. The hottest area is in the south-east, but the greatest number of fires occur in regions with lower temperatures. California covers an area of 423,970 km2 and has a very diverse topography (Wakabayashi & Sawyer, 2001). The Sierra Nevada, the fertile lands of the Central Valley and the arid Mojave Desert to the south are some of the geographical features of this North American state. It is home to some of the world's most exceptional trees that have not been affected by forest fires, the Sequoias, as well as the highest point, Mount Witney, and the lowest, Death Valley, in the 48 contiguous states. Figure 3 shows a representative flowchart of the impacts and changes in the socio-economic environment that occur in areas affected by forest fires. There is a reduction in recreational and leisure areas and economic damage caused by the destruction of forest stands.

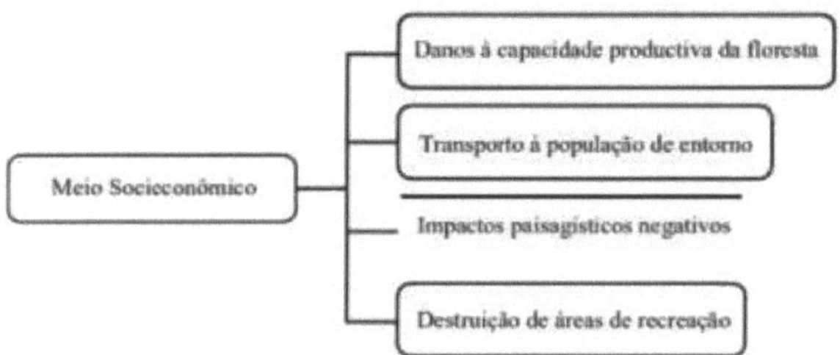

Figure 1 Flowchart of Fire Impacts on the Socio-Economic Environment
Source: Adapted from Guimaraes et al., (2014)

The fire also denudes the soil and leaves a blackish colour to the environment, which causes landscape monotony. This fact is also highlighted by Saldanha & Botelho (2008), who say that the aerosol particles emitted by fires, in addition to having an effect on the rainfall cycle, temperature, soil degradation, effects on biota and animal life, most probably also have adverse effects on the respiratory system of people who live in regions considered to be at risk, relatively closer to the areas that have been constantly burnt.

Table 3

Number of visitors/year to a national monument in the southern coastal zone of California. Source: Adapted from Dailey et al. (1974)

Table 3

Number of visitors /years for the national monument in the southern California coastal zone. Source: Adapted from Dailey et al. (1974)

National Monuments		Number of visitors
Channel Islands Reereagdo (Campers and Tourists)		36.700
Naturalists, Travels of the Three Camps		9.000
Cabrillo National Monument	1.341.200	
Recreation (Campers and Tourists) Naturalists, Country Journeys)		96.(XX)

Table 3 shows the number of visitors to California parks where fires have been a constant over the years. Tourism is intensely affected during the drier seasons, when forest fires always occur. The extent of the impacts caused by the fire's interaction with the physical environment affects water, soil and air. Water quality is affected by the concentration of ash carried away by the waters, as well as by the deposition of sediments detached from the soil as a result of erosion at the bottom of water bodies. Meixner (2004), after carrying out a study on the impacts of forest fires on areas immediately adjacent to the fires, concluded that, in terms of physical impacts on water, the increase in sediment flow after a fire has an impact on both ecology and drinking water, since water reservoirs and basins can be filled, disturbed and therefore damaged by sediment. Spencer, Gabel, & Hauerc. (2003) state in their study that one of the first impacts witnessed after a forest fire is the death of fish, which can be caused by high temperatures during the fire and/or the diffusion of smoke into the water, or by the alteration of water chemistry parameters, such as the high volatilisation of mobile compounds. Forest cover in the United States The predominant age of timber trees in the United States varies according to region. In the South, where more hectares of short-lived yellow pines are planted, 51 per cent of timberland is less than 40 years old, compared to 20 per cent in the North and 22 per cent in the West. In contrast, 56 per cent of logging areas in the North are over 60 years old, compared to 27 per cent in the South and 69 per cent in the West (Westober, 2016).

As for analysing the social space, Figure 2 shows land occupation between the urban and rural spaces that were not spared by forest fires and is methodologically and theoretically related to three general concepts: form, structure and 140

The formal and functional analyses do not eliminate the need to consider the scale, proportion, size and mobility of these occupied spaces.

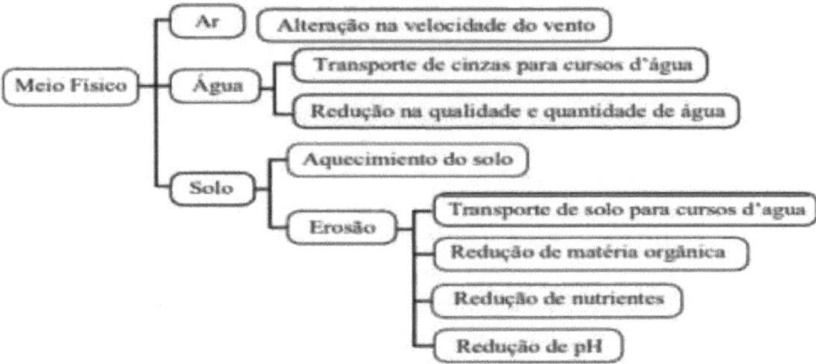

Figure 2 Flowchart of Impacts caused by Fire Interaction with the Physical Environment

Source: adapted from Guimaraes et al., (2014)

Results and discussion

When analysing forest fire occurrences, it's important to take into account the higher temperatures, especially in summer when fuel is more abundant, according to the data shown in figure 3. A higher incidence of fires in California occurs in the summer season, but throughout the year they have also been recorded in wetter and cooler seasons. Low humidity, according to Torres (2006), would lead to a deficit in cloud formation, increasing solar radiation on the surface. Temperature behaviour is one of the factors that must be taken into account when it comes to the occurrence and spread of forest fires. Larcher (2000) states that drier air can force the environment to undergo greater evaporation and plant transpiration, which ends up forming a greater amount of combustible material, facilitating the rapid spread of fire (Nobre, Sampaio, & Salazar, 2007). Figure 5 shows that between the months of May and August there is an increase in temperature in the State of California during the summer, probably increasing the risk of large forest fires which have devastated many rural and urban areas. With the summer season, the undergrowth remains dry, increasing its flammability, which leads to an increase in the number of fires through the rapid spread of the fire, making it more difficult to extinguish.

The average monthly temperatures seen in Table 4, according to the historical series of the State of California, could justify the fires between April and October, an interval in which the temperatures have always been higher, but what has caught the attention of fire experts is that fires have occurred in every month of the year, which represents an anomaly, since in other regions of the country, phenomena of this nature do not occur. According to Guimaraes et al. (2014), the presence of fire can cause damage and harm to populations indirectly linked to the area where fires occur, through excessive smoke, respiratory problems, air pollution from wind-borne particles, impaired visibility, landscape impacts and the destruction of recreational areas, preventing local tourism, which is highly abundant in Southern California. Wildfires impact the physical, biotic and socio-economic environments, producing effects on the ecosystem as a whole and initiating new impacts.

Table 4 Months with the highest temperatures in California

High	Jan	Feb	Mar	Apr	May	Jun	Jul	Aug	Sep	Oct	Nov	Dec
Los Angeles	68°F	68°F	70°F	72°F	74°F	78°F	83°F	84°F	83°F	78°F	72°F	67°F
San Francisco	57°F	60°F	63°F	65°F	67°F	69°F	70°F	70°F	72°F	70°F	63°F	57°F
Sacramento	55°F	61°F	66°F	73°F	81°F	89°F	93°F	93°F	88°F	78°F	64°F	55°F
Palm Springs	70°F	73°F	79°F	86°F	93°F	102°F	106°F	105°F	100°F	89°F	77°F	69°F
Eureka	56°F	56°F	56°F	57°F	59°F	61°F	62°F	63°F	64°F	62°F	58°F	55°F

Source: weatherspark.com

Changes to the biotic environment The biotic environment is made up of flora, fauna and micro-organisms. When out of control, fire, which is widely used to clean up the remains of agricultural crops, control weeds and renew pastures, can cause destruction of the surrounding crops and alter the local microclimate. The gaps left by the passage of fire favour the development of light-dependent species and, with this, a new plant succession emerges. Fire also acts on the arboreal extract, influencing, above all, the reduction in the size of trees and an increase in their crookedness, as well as the floristic and structural composition of the vegetation (Martins et al., 2009). The occurrence of fires in combination with other environmental changes attributed to man may have resulted in unforeseen consequences for natural ecosystems. Santos, Fernandes, Mafra, & Almeida (2003) evaluated the changes in the soil's chemical attributes resulting from the burning of native pastures. Maluleke & Mokwena (2017) consider the possibility of a relationship between climate change and the occurrence of fires. Some large forest fires in the United States have been blamed on years of fire suppression and the continued expansion of people into ecosystems where fire has been an almost constant presence. Each year, fire exposes larger portions of the population, making them vulnerable to its effects (Gongalves et al., 2012). Natural forest fires have always occurred as a normal element in the functioning of ecosystems. Zalzal (2018) states that forest fires are also considered to be one of the main drivers of plant evolution. Despite the destruction, fire allows the regeneration of various ecosystems and the production of a series of habitats in which different organisms can thrive (Reineking, Weibel, Conedera, & Bugmann. 2010). The enormous proliferation of fires caused by human activity in recent decades has exceeded the capacity for natural recovery, but hundreds of unexplained fires are still being investigated to find out the real causes. The behaviour and severity of a fire are the result of a combination of factors such as available fuels, hydraulic configuration and climate (Hutt, 2008). Analyses of historical meteorological data and national fire records in western North America show the primacy of climate in driving large regional fires during wet periods (Littell., et al. 2009). One of the most dangerous techniques used as an agricultural resource, very common in the Amazon and the Brazilian Midwest, is burning, which can often lead to a loss of control in the spread of fires (Fearnside, 2005). In California, this technique has been adopted because it does not incur many costs, but it is considered ecologically inappropriate for forest areas (Fidelis & Pivello, 2011). Fuels can also be removed by logging, but fuel treatments and thinning have no effect on severe fire behaviour in extreme weather conditions, making it very important for residents in high-mobility risk areas to protect their homes with fireproof materials. The

history of fires in California goes beyond human perception, mainly because the United States is a country with special conditions for surveillance, detection and mitigation in the event of natural disasters. As a wealthy country, spending on suppression doesn't put a strain on the government's finances, but the environment has been severely punished, including an increase in the destruction of structures and the number of victims in recent years. The number of acres burnt involves both wilderness areas and areas used for agriculture, causing great losses for the residents of the affected regions.
The disastrous effects of wildfires seem to be part of the scenario for residents of the State of California. The California Department of Forestry and Fire Protection (CDFFIRE), whose reference can be found in Table 3, is an agency responsible for 31 million acres of land, including public and private forests. Table 5 presents a historical overview of the State of California in relation to the occurrence and consequences of wildfires between 1999 and 2012. According to data obtained from fire.ca.gov, 71,083 fire calls were registered, with suppression costs totalling around 2,926,000, with the destruction of 13,845 structures including commercial establishments, homes, farms and a total of 2,488,644 acres totally destroyed by small, medium and large fires.

Table 5 Summary of California Incendies between 1999/2012

Year	№ of fires	Suppression costs	Destroyed Estnituras	Acres Quimados
1999	7.562	178 nulhoes	1.385	285.272
2000	5.177	114 nulhoes	130	72.718
2001	6.223	166 nulhoes	389	90.985
2002	5.759	135 nulhoes	327	112.810
2003	5.961	252 nulhoes	5.394	404.328
2004	5.574	170 nulhoes	1.016	168.134
2005	4.908	117 nulhoes	102	74.004
2006	4.805	206 nulhoes	431	222.896
2007	5.647	524 nulhoes	3.079	425.238
2008	4.923	500 nulhoes	992	347.810
2009	3.546	274 nulhoes	121	73.098
2010	2.961	51 nulhoes	94	23.191
2011	3.056	90 million	137	51.880
2012	4.981	140 million	248	136.280

Source: Adapted from CDF FIRE/www.fire.ca.gov/communications92018

For forest fire agencies, this represents a 21 per cent increase in the number of fires and a 38 per cent decrease compared to 2012. The Call Fire Agency recorded 6,196 fire incidents that burnt 123,053 acres, which represents a 20 per cent increase in the number of fires and an 11 per cent drop in the acres burnt in 2012. Forest fire agencies responded in 2014 to 7,233 calls to fight fires that burned 625,540 acres. For forest fire agencies, this represents a small decrease in the percentage of fires. In 2015, forest fire agencies responded to 8,283 calls that burned 880,899 acres. For forest fire agencies, this represents a 15 per cent increase in the number of fires and a 41 per cent increase in acres burned from 2014. During 2016, local, state and federal fire agencies responded to 6,954 wildfires that destroyed 669,534 acres throughout California. However, despite the increase in the number of fires compared to previous years, there was a greater destruction of hectares. One fact that has drawn the attention of fire investigation departments between 2013 and 2017 is the number of incidents whose cause is undetermined, especially in recent years. There has been a growing increase in undetermined causes, the figures provided by cdfdata.fire.ca.gov being 8 cases for 2013; 20 for 2014; 21 for 2015 and 29 undetermined causes for 2016. Forest fire statistics The National Interagency Fire Centre (NIFC) located in Boise-Idaho reported in its statistics that since 1960 there have been more than 60,000 fires in the United States each year. Lands that were considered wilderness are giving way to agricultural production, as well as an increase in the area of urbanisation, favouring an increase in population. Between

1960 and 2013, 5.6 million acres were burnt, according to NIFC data. In 54 years, the area burnt was 235.8 million acres, which is equivalent to 362,849 square miles. Some human activities are pointed out by the agencies as one of the main causes of fires, ranging from simple fires that have not been put out to intentional arson. Lightning is responsible for 14 per cent of the 823,032 forest fires that occurred between 2002 and 2012. Controlled burn-offs to establish agricultural and livestock areas can get out of hand and therefore account for 18 per cent of fires.

Final considerations

Whatever the cause of the forest fire, the consequences are always very serious, regardless of the extent of the fire. The vegetation affected can respond positively or negatively to the devastation caused by the fire. In some cases, plant species may adapt in an attempt to survive the chaos and regenerate in a short space of time, especially when there are no repeated fires in the same area. In higher risk areas, the analysis methods used to quantify and qualify the impacts often help to reduce the consequences of an unexpected disaster. A number of impacts are generated, including those on the human, biotic, social and economic environment, as well as the destruction of landscapes and recreational areas, which are very abundant in California. The development of housing in forested areas has increased the risk of fatal accidents, especially since 1989 when large fires became more frequent, affecting large contingents in the land-urban interface zone. Among the months of the year, the month of October presents the greatest possibility of events due to the winds, but in the southern part of California, the wind does not cause large fires and an example of this was the huge Zaca Fire, a large fire that burned for two months with only six days of strong winds. In recent years, California has been faced with a high number of fires that have drawn the attention of the authorities because of the devastation to homes near or not near hillsides whose dense forest prevents the incursion of agents to fight the spread of the fire. Another area that suffers from fire pollution are lakes and streams, whose fauna can be completely destroyed by water contamination. The toxic ash, in addition to the mountain of rubbish that should be placed in landfill sites, often invades sewage pipes and can contaminate water treatment plants. Forest fires are mainly responsible for destroying the electricity grid, complicating life and communication for residents.

References

Christensen, N.L., Bartuska, A.M., Brown, J.H., Carpenter, S., Francis, R., Franklin, J.F., ... Woodmansee, R.G. (1996). The report of the ecological society of America committee on the scientific basis for ecosystem management. Ecological application, 6(3), 665-691. https://doi. org/10.2307/2269460

Dailey, M. D., Reish, D. J., & Anderson, J. W. (1993). Ecology of the southern California bight. Los Angeles: University of California press. Fearnside, P.M. (2005). Deforestation in Brazilian Amazonia: history, rates, and consequences. Conservation biology, 19(3), 680688. https://doi.org/10.1111/j.1523-1739.2005.00697.x162

Fidelis &, A. & Pivello, V.R. (2011). Should fire be used as a management tool in the cerrado and southern grasslands? Brazilian Biodiversity, 2, 12-25. Retrieved from http://www.icmbio.gov.br/ revista eletronica/index.php/BioBR/article/ view/102/103

Galik, C.S. & Grala, R.K. (2017). Conservation programme delivery in the southern U.S: Preferences and interactions. Journal of environmental management, 198, 75-83. https://doi.org/10.1016/j. jenvman.2017.04.043

Goncalves, K.d.S., Castro H.A.de., Hacon, S.d.S. (2012). The fires in the Amazon region and respiratory illness. Ciencia e saude coletiva, 17(6), 1523- 1532. http://www.scielo.br/pdf/csc/v17n6/ v17n6a16.pdf

Guimaraes, P., de Souza, S., Fiedler, N., & da Silva, A. (2014). Analysing the environmental impacts of a forest fire. Agrarian academy, 1(1), 38-60. https://doi.org/10.18677/agrarian academy 2014 005

Horton, K., Knight, H., Galvin, K.A., Goldstein, J.H., & Herrington, J. (2017). An evaluation of landowners' conservation easements on their livelihoods and wellbeing. Biological conservation, 209, 62- 67.https://doi.org/10.1016/j.biocon.2017.02 .016

Hutt, R.L. (2008). The ecological importance of severe wildfires: some like it hot. Ecological applications, 18(8), 1,827- 1,834. https://doi.org/10.1890/08-0895.1

Larcher, W. (2000). Plant ecophysiology. Sao Carlos: RIMA. Artes e Textos. Littell, J.S., McKenhzie, D., Peterson, D.L., & Westerling, A.L. (2009). Climate and wilfire area burned in western U.S. ecoprovinces, 1916-2003. Ecological applications, 19(4), 1.003-1.021. https:// doi.org/10.1890/07-1183.1.

Lourengo, L. (1992). Assessment of fire risk in the woods and forests of mainland Portugal. Finisterra, 27(53-54), 115- 140. https://doi.org/10.18055/finis1883

Meixner, T. (2004). Wildfire impacts on water quality. Southwest hydrology, 3(5), 24-25. Retrieved from http://www.swhidro. arizona.edu/archive/V3_N5/feature7.pdf

Martins, V., Miranda, A.I., Carvalho, A., Schaap, M., Borrego, C. (2009) Impact of forest fires on air quality in Portugal from 2003 to 2005. Silva Lusitana, 17(2). Retrieved from http:// www.scielo.mec.pt/scielo.php?script=sci_ ext&pid=S0870- 63522009000200007

Maluleke, W. & Mokwena, R J. (2017) The effect of climate change on rural livestock farming: case study of Giyani Policing Area, Republic of South Africa. South African journal of agricultural extension, 45(1). http://dx.doi.org/10.17159/2413-3221/2017/ v45n1a404.

Nobre, A.C., Sampaio, G., & Salazar, L. (2007). Climate change and the Amazon. Ciencia e cultura, 59(3), 22-27.

Reineking, B., Weibel, P., Conedera, M., & Bugmann, H., (2010) Environmantal determinants of lightning-v.human-induced forest fire ignitions differ in a temperate. International journal of wildland fire, 19(5), 541-557. https://doi.org/10.1071/WF08206. 163 Environmental impacts caused by wildfires, California

Rheinheimer, D.S., Santos, J.C.P., Fernandes, V.B.B., Mafra, A.L., & Almeida, J.A. (2003). Changes in soil chemical attributes under native grassland subjected to burning. Ciencia rural, 33(1), 49-55. https://doi.org/10.1590/ s0103-84782003000100008

Saldanha, C.T. & Botelho, C. (2008). Burns and their influences on asthmatic children under

the age of five treated in a public hospital. Revista brasileira de alergia e inmuno patologia, 31(3), 108- 112.
Segerson, K. (2013). Voluntary approaches to environmental protection and resource management. Annual review of resource economics, 5(1),161- 180. https://doi.org/10.1146/annurev-resource091912-151945
Spencer, C.N., Gabel, K.O., Hauer, F. R. (2003). Wildfire effects on stream food webs and nutrient dynamics in glacier national park, USA. Forest ecology and management, 178(1 -2), 141-153. http:// doi.org/10.1016/S0378-1127(03)00058-6
Torres, F.T.P. (2006). Relationships between climatic factors and occurrences of forest fires in the city of Juiz de Fora (MG). Caminhos de geografia, 7(18), 162-171.
Wakabayashi, J.& Sawyer, T.L. (2001). Stream incision, tectonics, uplift, and evolution of topography of the Sierra Nevada, California. The journal of geology, 109(5), 539-562. https://doi. org/10.1086/321962
Westover, R. (2016, february) Western water threaten edby wildfire. [U.S. department of agriculture blog]. Retrieved from https://www. usda.gov/media/blog/2016/02/8/westernwater-threatened-wildfire.
Westerling, A L. (2006). Warming and earlier spring increase western U.S. forest wildfire activity. Science, 313(5789), 940- 943. https://doi.org/10.1126/science.1128834
Zalzal, K.S. (2018). A flammable planet: fire finds its place in Earth history. EARTH Magazine. Retrieved from https://www.earthmagazine.org/article/flam mable-planet-fire-finds-its-place-earthhistory.

Printed by Books on Demand GmbH, Norderstedt / Germany